Wideband FM
Techniques for Low-Power
Wireless Communications

RIVER PUBLISHERS SERIES IN CIRCUITS AND SYSTEMS

Volume 3

Series Editors

MASSIMO ALIOTO
National University of Singapore
Singapore

KOFI MAKINWA
Delft University of Technology
The Netherlands

DENNIS SYLVESTER
University of Michigan
USA

The "River Publishers Series in Circuits & Systems" is a series of comprehensive academic and professional books which focus on theory and applications of Circuit and Systems. This includes analog and digital integrated circuits, memory technologies, system-on-chip and processor design. The series also includes books on electronic design automation and design methodology, as well as computer aided design tools.

Books published in the series include research monographs, edited volumes, handbooks and textbooks. The books provide professionals, researchers, educators, and advanced students in the field with an invaluable insight into the latest research and developments.

Topics covered in the series include, but are by no means restricted to the following:

- Analog Integrated Circuits
- Digital Integrated Circuits
- Data Converters
- Processor Architecures
- System-on-Chip
- Memory Design
- Electronic Design Automation

For a list of other books in this series, visit www.riverpublishers.com

Wideband FM
Techniques for Low-Power
Wireless Communications

John F.M. Gerrits

CSEM, Switzerland
and
Delft University of Technology, The Netherlands

River Publishers

Published, sold and distributed by:
River Publishers
Alsbjergvej 10
9260 Gistrup
Denmark

River Publishers
Lange Geer 44
2611 PW Delft
The Netherlands

Tel.: +45369953197
www.riverpublishers.com

ISBN: 978-87-93379-62-6 (Hardback)
    978-87-93379-61-9 (Ebook)

©2016 River Publishers

Hold on tight to your dream

Contents

Foreword: John R. Long

This book outlines a low-power communication technology from system to circuit levels, including results obtained from three generations of hardware prototypes. It is mainly the work of a single person, author John F. M. Gerrits. John was a keen inventor, but primarily an engineer who was dedicated to his craft in a way that I have rarely seen. His vision for energy-efficient, robust and license-free communication led him to propose the concept of wideband- and ultrawideband-FM (UWB-FM), which relies upon a double-dose of frequency modulation to simplify the wireless transceiver. The results of Gerrit's work developing UWB-FM are documented in this small volume. As you will discover in the following pages, UWB-FM is remarkable in that a phase-locked loop is not required in either the radio transmitter or receiver. This simplification drastically reduces power consumed during the start-up and shut-down of a conventional transceiver, which is important in low-power applications where duty cycling of the transceiver power supply is necessary. By contrast, energy consumed by a UWB-FM transceiver is directed towards the communication of information, and not wasted in the production of a stable frequency reference. The choice of FM, rather than the energy-efficient AM often seen in ultra-low-power "wake-up radio" prototypes, gives the UWB-FM transceiver greater robustness to interference caused by other wireless appliances in the home or work environments. The principle that guided Gerrits in the development of UWB-FM was elegance and simplicity. He believed that realizing both would result in a technology that could be inexpensive, reliable and green; a combination of attributes that has eluded many other ultra-low-power radio designers.

John created a unique technology with the ultrawideband-FM concept, and he brought it to light through hard work, dedication and very close attention to detail. Gerrits relied on simulations extensively, but he was also a highly-skilled experimentalist who constructed high-performance proof-of-concept prototypes from low-cost, discrete components. He was a key figure in the European projects MAGNET and MAGNET-Beyond, which drew researchers from European companies, research institutes and universities into the pursuit of a more complete UWB-FM technology that included a protocol and application benchmarks. John Gerrits was also motivated to pursue a Ph.D. degree based on his research findings, which is the source of the material for this book. While the story for Mr. Gerrits personally ended through his own hand in February of 2011, that was not the final act for UWB-FM. The value of his ideas inspired others to join in the quest to build a low-power wideband FM transceiver on a silicon chip. A complete, single-chip UWB-FM transceiver in silicon-CMOS technology was demonstrated in 2014 by Saputra [1], and a receiver by Kopta [2] will be presented in 2016. Indeed, UWB-FM is poised to play a role in the development of radio technologies for the emerging Internet of Things. Hold on tight to that dream!

John R. Long
Waterloo, Ontario
April 24, 2016

References

[1] N. Saputra and J. R. Long, "A Fully Integrated Wideband FM Transceiver for Low Data Rate Autonomous Systems," *IEEE Journal of Solid-State Circuits Special Issue on the 2014 RFIC Symposium*, vol. 50, no. 5, pp. 1165–1176, May 2015.

[2] V. Kopta, D. Barras and C. C. Enz, "A 420-μW, 4-GHz approximate zero-IF FM-UWB receiver for short-range communications," IEEE-RFIC Symposium, San Francisco CA, May 2016.

Foreword: Cees J. M. Lanting

For John – John F. M. Gerrits – there were three things of importance in life: technology, his wife Aimée, and friendship.

John dedicated an important part of his being to technology.

John was constantly busy with his research and development work, almost 24 hours and 7 days per week, and even for Aimée it was sometimes hard competing; it could be better, had to be better, there were certainly other, better solutions, ...

John and I frequently discussed ideas and results, and it was a privilege for me to be able to help him with comments and ideas, or better, help him to find the ideas himself.

John simply loved technology. In the Magnet-Beyond project he made some good quality pictures of special designed semiconductors that had been developed for the project: these were so good looking, art and technology . . . 'For an aircraft to fly well, it must be beautiful', plane designer and manufacturer Marcel Dassault used to say. And for John, this principle had wider application.

Later, when the designers of the 'chips' learned about the photos, they got suspicious and thought 'reverse engineering' had been attempted. But John's photos were not made for that purpose. And, reverse engineering would not even be considered in John's high standard of ethics.

When by accident I told John that the wireless internet access in my apartment was not so good, John immediately arrived with a set of measurement tools: this was a problem that had to be solved! And we found the best location for the gateway (outside a window), and some 'tricks' to improve the signal propagation.

In the last weeks of his life, John felt increasingly under pressure: he was under the impression that his projects and his work were less and less appreciated, and even felt that what for him doing the right thing, was considered waste of time and less useful by others: his technology world was risking to fall apart, he feared.

Aimée and John had a fantastic relationship, partially based on how different they were. But so different, they sometimes could not understand the other's problems and help. And one day it all seemed to be too much to John, not possible to overcome it seemed . . .

And we lost an excellent scientist, a dedicated husband and companion, and a great friend.

Thanks to the promoter, Prof. John R. Long, and the significant effort and great tenacity he put in, this book has become ready for publication.

This book is part of the legacy of John F. M. Gerrits' work, making this part of this life more visible, and hopefully will bring the recognition that it and he deserve.

Cees J. M. Lanting
Bern, Switzerland
May 09, 2016

Acknowledgments

I am greatly indebted to my promotor Prof. Dr. J. R. Long for the guidance he has provided throughout the book and the helpful comments on the manuscript of this book.

I would like to thank CSEM's Human Resources Department and Management, who granted me favorable conditions for writing the book, and especially Dr. John Farserotu, who has made it possible to work out the FM-UWB concept in the European research projects MAGNET and MAGNET Beyond. Many thanks to the European Commission for sponsoring our work and all the research partners who helped to show the feasibility of the FM-UWB approach.

I would like to thank Gerrit van Veenendaal and Hamid Bonakdar from NXP Semiconductors for their invaluable support in realizing the high band receiver front-end and subcarrier processing. Many thanks to Yi Zhao and Yunzhi Dong from TU Delft who made the first IC realizations of the high band FM-UWB receiver front-end. Further thanks go to Manuel Lobeira, Marco Detratti and Ernesto Perez from ACORDE S. A. for their implementation of the high band RF VCO and output amplifier, and to Peter Nilsson and Jiren Yuan from Lund University for the implementation of the DDS IC. Thanks to all Magnet Beyond partners who helped to make the FM-UWB prototypes a success.

Thanks go also to the Staff of the Electronics Research Laboratory for their hospitality and the pleasant working atmosphere that has always been around for their "Man from Switzerland".

I am also very grateful to my parents who have always encouraged me to become an electrical engineer and make my dreams come true. Last, but not least, my gratitude to Aimée, for her patience and understanding.

List of Figures

List of Tables

List of Abbreviations

ADC	Analog-to-digital converter
AGC	Automatic gain control
AM	Amplitude modulation
AWGN	Additive white Gaussian noise
BER	Bit error rate
CDF	Cumulative density function
DAA	Detect and avoid
DAC	Digital-to-analog converter
DDS	Direct digital synthesis
DFT	Discrete Fourier transform
EIRP	Equivalent isotropic radiated power
ETSI	European Telecommunications Standards Institute
FCC	Federal Communications Commission
FDMA	Frequency division multiple access
FLL	Frequency-locked loop
FM	Frequency modulation
FSK	Frequency shift keying
IR	Impulse radio
kbps	kilobit per second
LDR	Low data rate
LNA	Low noise amplifier
LO	Local oscillator
LOS	Line of sight
MA	Multiple access
MAC	Multiple access control
MBOFDM	Multi-band orthogonal frequency division multiplexing
Mbps	Megabit per second
MDR	Medium data rate
OA	Output amplifier
OFDM	Orthogonal frequency division multiplexing
PAM	Pulse amplitude modulation

PDF	Probability density function
OOK	On-off keying
PL	Path loss
PM	Phase modulation
PLL	Phase-locked loop
PPM	Pulse position modulation
Q	Quality factor
QPSK	Quaternary phase shift keying
RF	Radio frequency
RMS	Root mean square
SIR	Signal-to-interference ratio
SNR	Signal-to-noise ratio
TFC	Time-frequency coding
TFI	Time-frequency interleaving
TH	Time hopping
UWB	Ultra-wideband
VCO	Voltage controlled oscillator
WiMAX	Worldwide interoperability for microwave access

List of Symbols

A	Amplitude
B_{DEM}	Demodulator bandwidth
B_{RF}	Rf bandwidth (of FM-UWB signal)
B_{SUB}	Subcarrier bandwidth
Cf	Crest factor
C_{MAX}	Maximum capacity
d	Data signal
d_{FS}	Distanced covered under free space propagation conditions
E_b	Energy per bit
F	Noise factor
f_C	Center frequency
f_m	Modulation frequency
f_{CLK}	Clock frequency (of DDS)
f_{SUB}	Subcarrier frequency
G_n	n^{th} derivative of the Gausssian wavelet
G_P	Receiver processing gain
G_{TX}	Transmit antenna gain
G_{RX}	Receive antenna gain
h	Channel time domain impulse response
H	Channel frequency domain transfer function
HD_1	Amplitude of first harmonic
HD_n	Amplitude of nth harmonic
H_{LPF}	Transfer function of lowpas filter
K_D	FM demodulator sensitivity
k	Boltzmann's constant
k_g	Gain distribution factor
L(f)	Single-sided power spectral density
m	Subcarrier signal
M	Subcarrier waveform (sine, triangle, sawtooth)
N	Number of users
N_{MAX}	Maximum number of users

NF	Noise figure
N_0	Noise power spectral density
O	Overdrive
O_{dp}	Overdrive of differential pair
O_{quad}	Overdrive of multiplier quad
o	Relative frequency offset
q	Elementary charge
P_{AV}	Average power
PC	Channel periodicity
P_{NRX}	Equivalent input noise power
P_{RX}	Received power
P_{TX}	Transmitted power
p_b	Probability of error
P_{PEAK}	Peak power
R	Transmission rate
S_F	Power spectral density of random frequency fluctuations
S_P	Power spectral density of random phase fluctuations
t	Time
T	Period time, or absolute temperature
V	UWB signal
V_{DEM}	Demodulator output signal
V_{FMDEM}	Transfer function of FM demodulator

Greek

Δf	FM-UWB deviation
Δf_{SUB}	Subcarrier deviation
α	Ratio of collector current to emitter current
β	FM-UWB modulation index
β_{SUB}	Subcarrier modulation index
δ	Duty cycle
δf	Frequency resolution (of DDS)
ϕ	DDS phase

1

Introduction

Ultra Wideband (UWB) communications are poised to enable short-range applications, such as remote health monitoring (e-health) and home or office automation [1]. Sensor networks are also suitable candidates for UWB since the low radiated power of the UWB transmitter enables low DC power consumption, yielding long battery life and the possibility to use energy scavenging. Size and cost constraints require a low-complexity approach that allows multiple users to share the same RF bandwidth, and offers robustness to interference, frequency-selective multipath and antenna mismatch.

1.1 Definition of a UWB Signal

By definition [2], the –10 dB RF bandwidth (B_{RF}) of a UWB signal should be at least 20% of the center frequency f_c when f_c is less than 2.5 GHz and at least 500 MHz when f_c is greater than 2.5 GHz.

Apart from the bandwidth constraint, UWB signals must meet the emission spectral mask as defined by the local regulatory authorities [2, 3]. Restrictions apply to the average power spectral density and the peak power of the signal radiated by the transmitter. For a UWB transmitter operating between 3.1 and 10.6 GHz, the average equivalent isotropic radiated power (EIRP) is limited to –41.3 dBm/MHz, whereas the peak EIRP should not exceed 0 dBm in a 50 MHz bandwidth [2]. A UWB transmitter with 500 MHz bandwidth has its average output power limited to –14.3 dBm and its peak power to 10 dBm. The exact relationship between average and peak power is determined by the envelope of the signal. For a constant-envelope signal, average and peak power are equal, whereas for a sine wave burst with duty cycle (δ) and pulse envelope with crest factor (Cf), i.e., the ratio of peak and RMS amplitude, the relation is given by

$$P_{AV} = \frac{\delta}{Cf} P_{PEAK}. \tag{1.1}$$

This ultra-wide bandwidth opens the door to high capacity wireless communication systems, where data rates of hundreds of Mbps are possible. Alternatively, large numbers of low data rate (LDR; data rate less than 100 kbps) and medium data rate (MDR; data rate between 100 kbps and 1 Mbps) users and their devices can be supported. Apart from the advantage of unlicensed worldwide operation, UWB technology brings robustness to multipath fading [4], good coexistence with other communication systems due to its low power spectral density and low transmit power. The wide bandwidth also offers the possibility to perform accurate ranging [5]. Sensors in a sensor network can determine their relative position which is required for the implementation of efficient multi-hop routing. In principle, the low radiated power of a UWB transmitter opens the door to low power consumption, which may be achieved using one, or several of the following techniques [6]:

- duty-cycling of transmitter and receiver electronics,
- simple radio architecture,
- low supply voltage/current (clever biasing techniques, current reuse),
- use of "narrowband" circuit techniques, i.e., limiting the receiver bandwidth to the bandwidth of the received signal.

In subsequent sections of this chapter, these techniques will be elaborated.

Since the definition of a UWB signal does not specify a particular air interface or modulation scheme, many different techniques yield a UWB signal. Various UWB approaches are either in-use today or are being studied, addressing applications such as position localization [5], high data rate transmission, or robust communication [7]. The next sections will present today's state-of-the-art UWB systems and the motivation for developing the constant-envelope FM-UWB approach.

1.2 Impulse Radio

Originally, UWB systems used a time-domain approach where a pulse generator generates baseband wavelets (monocycles) [8, 9] that are directly fed into the transmit antenna. Gaussian pulses and their derivatives are often used as wavelets in impulse radio systems. The Gaussian wavelet $g_0(t)$ is defined as

$$g_0\left(t\right) = \frac{1}{\sigma\sqrt{2\pi}}\exp\left(-\frac{t^2}{2\sigma^2}\right). \qquad (1.2)$$

The nth derivative $g_n(t)$ is defined as

$$g_n(t) = \frac{d^n}{dt^n} g_0(t). \qquad (1.3)$$

The pulse duration τ is approximately 6 σ. Figure 1.1a shows the Gaussian wavelet $g_0(t)$ and its first and second derivatives $g_1(t)$ and $g_2(t)$ for $\sigma = 100$ ps. The wavelets have been normalized to unity amplitude. Figure 1.1b shows the frequency spectrum of the wavelets. The Gaussian wavelet $g_0(t)$ has a DC component and a lowpass spectrum. This wavelet cannot be radiated

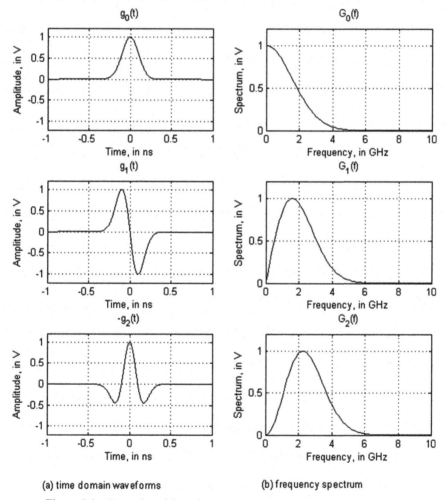

(a) time domain waveforms   (b) frequency spectrum

Figure 1.1 Examples of Gaussian wavelets in time and frequency domain.

efficiently by the antenna which has either a highpass or bandpass transfer function. Derivatives of the Gaussian wavelet work better since they show a bandpass spectrum. Various wavelet generation circuits approximating the Gaussian waveform have been published [10, 11].

For Gaussian pulses, the parameter σ defines both the center frequency and bandwidth. Because of the dependency between these two parameters, it is difficult to respect the emission mask as defined by the ETSI and the FCC [2, 3]. Variation of pulse duration, using derivatives of the Gaussian pulse and combination of pulses are theoretical ways to improve spectral efficiency [12, 13]. However, the slow spectral roll-off of the Gaussian pulse makes it difficult to implement a compliant impulse radio transmitter without additional pulse shaping and filtering. As a last resort, lowering the transmit power may be required to comply with the UWB emission mask.

In practical UWB systems, it is necessary to decouple pulse duration and center frequency. A UWB signal at center frequency f_c may be obtained by up-conversion of the baseband signal envelope. This carrier-based approach yields reproducible results. The baseband envelope is relatively easy to implement with today's IC electronics. Half-cosine and triangular waveforms make suitable candidates for the pulse envelope [14], and triangular waves are straightforward to generate using integrated circuit techniques [15]. Figure 1.2 shows time domain waveforms for carrier-based UWB signals with half-cosine envelope (waveform $p_1(t)$ with spectrum $P_1(f)$) and triangular envelope (waveform $p_2(t)$ with spectrum $P_2(f)$). The carrier frequency is 7 GHz, and the pulse duration assumed is 2 ns.

Figure 1.3 shows the block diagram of an IR UWB transmitter according to the IEEE802.15.4a standard [16]. Figure 1.4 shows the symbol structure used in this standard. The basic modulation scheme is pulse position modulation (PPM) with additional time hopping (TH) for spectral smoothing and multiple access (MA). This system requires close time synchronization between transmitter and receiver. Time offset should be typically less than the pulse duration to avoid performance degradation [17]. The finite accuracy of the reference clocks in both transmitter and receiver, typically on the order of ±100 ppm for a low-cost quartz and ±1% for a quartz-less reference oscillator, makes the implementation of the synchronization problematic.

The initial synchronization after a cold start is composed of frequency, frame and symbol synchronization [18]. Frequency synchronization is a prerequisite for the coherent combining of multiple received pulses in a burst. It is accomplished by measuring and adjusting the frame duration. Next, the remaining time delay between the transmitter and receiver is compensated

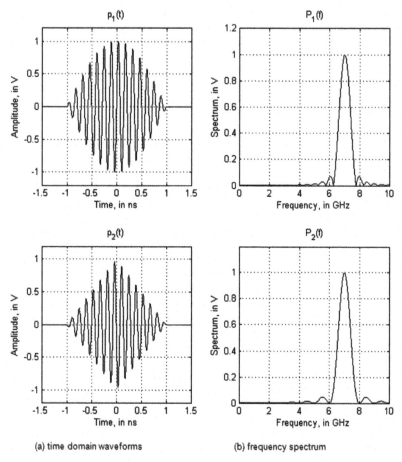

Figure 1.2 Examples of carrier-based IR UWB waveforms and their spectra [13].

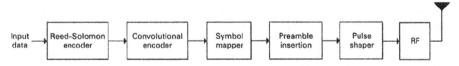

Figure 1.3 IR-UWB transmitter following the IEEE802.15.4a standard [16].

for. Frame synchronization compensates delays shorter than the frame time τ_f, and is successful when the moment of the pulse reception is known, i.e., the pulses are received within the acquisition windows. There may still be ambiguity in the symbol timing (see Figure 1.4). Symbol synchronization, which searches for the beginning of the spreading code, takes care of the

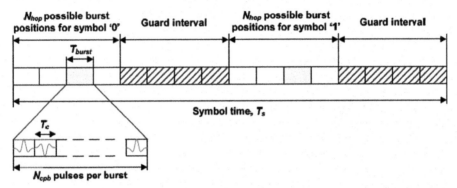

Figure 1.4   Symbol structure of IEEE802.15.4a impulse radio signal [16].

fine synchronization. In conclusion, synchronization to a low duty cycle pulse sequence is not a trivial matter and considerable effort needs to be invested in the synchronization hardware [19]. Reported power consumption of the digital back-end of the IR receiver is on the order of 2 mW for a 2.67 Mbps system [20]. For comparison, in a narrowband radio using on-off keying (OOK) or frequency shift keying (FSK), only bit synchronization is required.

The impulse radio receiver can be either coherent (correlation receiver) or non-coherent (energy detection receiver). Figure 1.5 shows the block diagram of a non-coherent receiver with low noise amplifier (LNA), square-law detector, lowpass filter (or integrator), sample-and-hold (S&H) and decision circuits. Synchronization logic provides the control signals for the integrator, sampler and decision circuit.

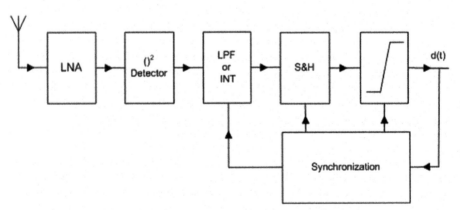

Figure 1.5   Non-coherent impulse radio receiver.

The low power consumption potential of impulse radio circuit implementations has been recently demonstrated. The transmitter reported in [15] operates from a 1 V supply and is based upon a digitally-controlled ring oscillator with well-defined start-up conditions and whose frequency is calibrated regularly using a frequency-locked loop (FLL). Required frequency accuracy at RF is 4 MHz. Most of the electronic circuits are active only for a brief time interval around the transmit pulse, and remain in standby mode for the rest of the time. With start-up and shut-down times in the order of 2 ns, these circuits allow the use of aggressive duty cycling to lower the average power consumption. Transmitter power consumption is 1 mW at a data rate of 1 Mbps, which corresponds to 1 nJ/bit.

The receiver front-end presented in [21] operates from a 0.65 V supply and uses a simple straight-through architecture with cascaded tunable RF bandpass filters followed by an envelope detector as shown in Figure 1.6.

Power is saved by adjusting the receiver bandwidth to the bandwidth of the received signal (typically 500 MHz). Filtering also creates frequency selectivity to reduce the effect of out-of-band interference. Instantaneous power consumption of this receiver front-end is 36 mW at a maximum data rate of 16.7 Mbps, which corresponds to 2.2 nJ per bit. Duty cycling is used to lower the power consumption at lower data rates.

At a first glance, duty cycling may seem to be an elegant approach to obtain low power consumption. However, it implies a complex control system for pulse synchronization [20].

Robustness is obtained from the high peak transmit power which results in high SNR at the receiver input. The use of short duration pulse results in fine multipath resolution capability [9]. Real processing gain as present in a direct

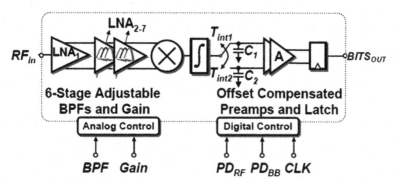

Figure 1.6 Impulse radio receiver block diagram [21].

sequence spread-spectrum system is not available. Ideally, robustness would
be implemented by a spread spectrum system with instantaneous despreading
(i.e., no pulse or sequence synchronization).

1.3 Wimedia

Today's standard for high data-rate UWB is the multi-band orthogonal frequency division multiplexing (MB-OFDM) approach, or WiMedia [7, 22],
which provides a robust solution for bit rates up to 480 Mbps. The ultra-
wideband OFDM signal consists of 128 subcarriers, one hundred of which
are modulated using QPSK techniques.

Figure 1.7 shows a block diagram of a MB-OFDM transmitter [23].
The baseband OFDM signal is digitally generated using IFFT techniques
and upconverted to the RF center frequency. The multi-band approach uses
multiple 528 MHz wide subbands. Center frequency for the nth subband, $f_c(n)$,
is given by

$$f_c(n) = 2904 + 528n, \text{ in MHz.} \tag{1.4}$$

The system may utilize time-frequency coding (TFC), also called time-
frequency interleaving (TFI), to interleave data over m subbands. This
increases transmission power by a factor of m compared to a single band
radio. A guard time (9.47 ns) provides time for transmitter and receiver
to switch to the next carrier frequency. First generation devices use three
subbands as shown in Figure 1.8 with center frequencies of 3432, 3960 and
4488 MHz. This approach requires a fast hopping frequency synthesizer with
phase continuity. It is required that the phase of the transmitted carriers be
coherent from hop to hop. Whenever the center frequency hops away from a
given frequency, the phase must be the same when it returns to the starting
frequency. This is to avoid the necessity for re-computing phase estimates on
each hop [7].

OFDM systems are designed in such a way that the bandwidth of a single
subcarrier is lower than the channel's coherence bandwidth. As a result,
the channel transfer function is flat over the bandwidth of each subcarrier.

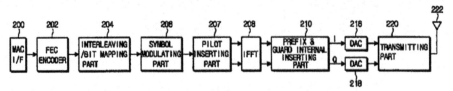

Figure 1.7   OFDM transmitter block diagram [23].

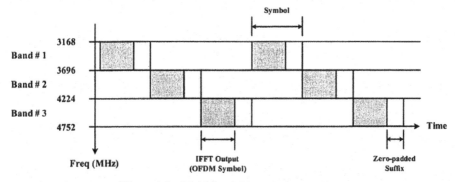

Figure 1.8 Multi-band OFDM concept [22].

No channel equalization is required. However, some subcarriers may have insufficient signal-to-noise ratio (SNR) for correct demodulation. This is overcome by coding and interleaving of the transmitted data.

This high-performance system comes at the expense of circuit complexity and higher overall power consumption, typically on the order of 200 mW for the radio front-end [24, 25]. However, given the high data rate, these solutions perform well in terms of energy per transmitted bit. A power consumption of 200 mW at 480 Mbps corresponds to 0.42 nJ/bit.

1.4 Motivation for Frequency Modulation FM-UWB

Figure 1.9 shows data rate, power consumption and energy per bit for state-of-the-art radios [1].

New wireless applications such as health monitoring and body-area networks (BAN) require tetherless connectivity at data rates below 250 kbps, a range less than 10 m, and operational lifetime from a single battery charge of weeks or months.

The IEEE802.15 Task Group 6 (IEEE802.15.6) is currently developing a communication standard for body-area networks [26] to address these low data rate (LDR) applications. The LDR radio for such applications needs to be low-complexity and yet robust to interference and frequency-selective multipath and also be able to rapidly join or leave a network. The short PHY synchronization time simplifies the task of the medium access control (MAC), which would then be similar to the ones used in low-power narrowband, but not necessarily robust, radios. Also, the use of duty-cycling techniques should be avoided for the sake of low complexity.

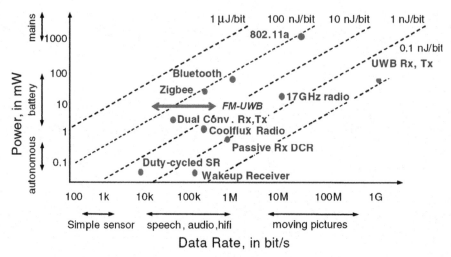

Figure 1.9   Energy consumption space for low power radio [1].

The challenge for the UWB radio designer can be stated as implementing a low-complexity, compact, low-power spread spectrum radio with instantaneous despreading in the receiver (no pulse or sequence synchronization).

Constant-envelope frequency modulation ultra-wideband (FM-UWB) addresses these low data (LDR) applications using a low-complexity implementation based upon a low-frequency FSK subcarrier followed by wideband analog FM to implement spread spectrum. Instantaneous despreading in the receiver is realized by an analog FM demodulator. The FSK subcarrier is demodulated in a classical way, i.e., using a direct conversion architecture. Bandwidth reduction after the FM demodulation yields the receiver processing gain which is given by the ratio of the RF and subcarrier bandwidth. As will be shown in the subsequent chapters, the ultra-wide bandwidth provides robustness to frequency selective multipath, whereas the processing gain provides robustness against interference. FM-UWB is competitive in terms of power consumption and has been proposed as one of the candidate air interfaces for medical body area networks [27].

1.5 Objectives and Scope of the Book

This book describes the principles of operation of the FM-UWB communication system and analyzes its performance for various cases. System level design of transmitter and receiver is presented. The book also considers practical implementation of the FM-UWB radio. FM-UWB is not just a theoretical concept; circuit level implementations of individual blocks and also the complete transceiver have been built to verify theory and to demonstrate the low power consumption potential of FM-UWB technology [28, 29]. These implementations were partly undertaken in the context of the European IST research projects Magnet and Magnet Beyond [30], addressing wireless personal area networks (WPANs). The lessons learned during the implementation of the FM-UWB radio are also part of this book.

1.6 Organization of the Book

The book is organized in the following way. Chapter 2 analyzes the theoretical principles of FM-UWB and gives system level design considerations for both transmitter and receiver. The optimum subcarrier waveform, frequency and modulation scheme are identified. Performance limiting factors, both fundamental and implementation-dependent are identified and discussed.

Chapter 3 analyzes the FM-UWB radio performance for various cases: additive white Gaussian noise (AWGN), frequency selective multipath, multiple users and interference.

Chapter 4 presents details on the transmitter implementation. It also addresses the effect of imperfections like VCO tuning curve non-linearity and oscillator phase noise on the transmitter performance.

Chapter 5 considers details of the receiver implementation. Its main focus is on the implementation of the wideband FM demodulator, key building block of the receiver. Implementation examples of the delay and multiplier circuits in this demodulator are given and the noise and large-signal behavior of a practical demodulator implementation is analyzed. The implementation of the subcarrier processing is also shown.

Chapter 6 presents performance as measured on a hardware prototype operating at 7.45 GHz. A comparison is made between the measured performance and the theory shown Chapter 2.

Chapter 7 closes the book with the main findings, a list of original contributions, and recommendations for further research.

References

[1] J. R. Long, W. Wu, Y. Dong, Y. Zhao, M. A. T. Sanduleanu, J. F. M. Gerrits, and G. van Veenendaal, "Energy-efficient wireless front-end concepts for ultra lower power radio," in *Proc. IEEE CICC2008*, pp. 587–590.

[2] Federal Communications Commission (FCC), *Revision of part 15 of the commission's rules regarding ultra wideband transmission systems*, "First Report and Order, ET Docket 98–153, FCC 02–48," Adopted: February 2002; Released: April 2002.

[3] ECC Electronic Communications Committee, *ECC Decision of 24 March 2006 on the harmonised conditions for devices using Ultra-Wideband (UWB) technology in bands below 10.6 GHz*, ECC/DEC/(06)04, APT/AWF (Amended ECC/DEC/(06)04), [Online]. Available: http://www.apwpt.org/downloads/eccdec0604uwb.pdf

[4] M. Wolf, G. Del Galdo, M. Haardt, "Performance evaluation of ultra-wide band compared to wireless infrared communications," in *Proc. ECWT2005*, pp. 193–196.

[5] *Sapphire DART, Ultra Wideband Real-Time Location System Datasheet*, [Online]. Available: http://www.multispectral.com/pdf/Sapphire_DART.pdf

[6] J.F.M. Gerrits, J.R. Farserotu, and J.R. Long, "Low-Complexity Ultra Wideband Communications," in *Proc. ISCAS 2007*, pp. 757–760.

[7] G. Heidari, *WiMedia UWB: Technology of Choice for Wireless USB and Bluetooth*, New York: Wiley, 2008.

[8] M.Z. Win, R.A. Scholtz, "Comparisons of analog and digital impulse radio for wireless multiple-access communications," in *Proc. ICC97*, pp. 91–95.

[9] M. Z. Win, and R.A. Scholtz, "Impulse Radio: How It Works," *IEEE Communications Letters*, vol. 1998, no. 2, pp. 36–38, February 1998.

[10] J.F.M. Gerrits and J.R. Farserotu, "Wavelet generation circuit for UWB impulse radio applications," *Electronics Letters*, vol. 38, no. 25, pp. 1737–1738, 5 December 2002.

[11] S. Bagga, W.A. Serdijn, and J.R. Long, "A PPM Gaussian monocycle transmitter for ultra-wideband communications," in *Proc. UWBST & IWUWBS 2004*, pp. 130–134.

[12] M.-G. Di Benedetto, and B.R. Vojcic, "Ultra Wide Band (UWB) Wireless Communications: A Tutorial," *Journal of Communication and Networks*, vol. 5, no. 4, pp. 290–302, December 2003.

[13] Tian Tong, Torben Larsen, "Concept and Architecture of an Integral Receiver for a Low Data Rate Ultra-Wide Band System," in *Proc. 1st International MAGNET Workshop*, November 2004.

[14] Jaouhar Ayadi, John Gerrits, Patrick Vaney, Andreas A. Hutter, and John Farserotu, "System Design and Performance Analysis of an Ultra-wideband based Low Data Rate Wireless Personal Area Network," in *Proc. HET-NETs 2003*, pp. 81/1–81/8.

[15] Julien Ryckaert1, Geert Van der Plas, Vincent De Heyn, Claude Desset, Geert Vanwijnsberghe, Bart Van Poucke, Jan Craninckx, "A 0.65-to-1.4nJ/burst 3-to-10 GHz UWB Digital TX in 90 nm CMOS for IEEE 802.15.4a," in *Proc. ISSCC2007*, pp. 120–121.

[16] IEEEP802.15.4a/D4 (Amendment of IEEE Std 802.15.4), Part 15.4: *Wireless medium access control (MAC) and physical layer (PHY) specifications for low-rate wireless personal area networks (LRWPANs)*, July 2006.

[17] Ramin Miri, Lei Zhou, and Payam Heydari, "Timing synchronization in impulse-radio UWB: Trends and challenges," in *Proc. NEWCAS-TAISA 2008*, pp. 221–224.

[18] Roman Merz, *Analysis of Low Power, Low Data Rate Ultra Wideband Impulse Radio Systems*, PhD Thesis, Université de Neuchâtel, February 2009.

[19] Marian Verhelst, Julien Ryckaert, Yves Vanderperren, and Wim Dehaene, "A low power, reconfigurable IR-UWB system," in *Proc. ICC2008*, pp. 3770–3774.

[20] Marian Verhelst, and Wim Dehaene, "A Flexible, Ultra-Low Power 35pJ/pulse Digital Back-end for a QAC UWB Receiver," in *Proc. ESSCIRC 2007*, pp. 236–239.

[21] F.S. Lee, and A. P. Chandrakasan, "A 2.5 nJ/b 0.65V 3-to-5GHz Subbanded UWB Receiver in 90nm CMOS," in *Proc. ISSCC 2007*, pp. 116–118.

[22] MultiBand OFDM Alliance, *MultiBand OFDM Physical Layer Specification*, Release 1.0, April 27, 2005. [Online]. Available: http://my.com.nthu.edu.tw/~jmwu/LAB/MB-OFDM_Spec_2005.pdf

[23] Torbjorn A. Larsson, Nishant Kumar, *MB-OFDM transmitter and receiver and signal processing method thereof*, US Patent 7,577,160, August 18, 2009.

[24] Raf Roovers, Domine M.W. Leenaerts, Jos Bergervoet, Kundur S. Harish, Remco C.H. van de Beek, Gerard van der Weide, Helen Waite, Yifeng Zhang, Sudhir Aggarwal, and Charles Razzell, "An Interferenc-Robust Receiver for Ultra-Wideband Radio in Si-Ge BiCMOS Technology," *IEEE Journal of Solid-Sate Circuits*, vol. 40, no. 12, pp. 2563–2572, December 2005.

[25] Christoph Sandner, Sven Derksen, Dieter Draxelmayr, Member, Staffan Ek, Voicu Filimon, Graham Leach, Stefano Marsili, Denis Matveev, Koen L. R. Mertens, Florian Michl, Hermann Paule, Manfred Punzenberger, Christian Reindl, Raffaele Salerno, Marc Tiebout, Andreas Wiesbauer,

Member, Ian Winter, and Zisan Zhang, "A WiMedia/MBOA-Compliant CMOS RF Transceiver for UWB," *IEEE Journal of Solid-Sate Circuits*, vol. 41, no. 12, pp. 2787–2794, December 2006.

[26] IEEE 802.15 WPAN^{TM} Task Group 6 Body Area Networks (BAN), website URL: http://www.ieee802.org/15/pub/TG6.html

[27] John Farserotu, John Gerrits, Jérôme Rousselot, Gerrit van Veenendaal, Manuel Lobeira, John Long, *CSEM FM-UWB Proposal*, IEEE P802.15 Working Group for Wireless Personal Area Networks (WPANs), Task Group TG6 Body Area Networks (BAN), May 2009, Montreal. Doc.: IEEE802.15-09-0276-00-0006, [Online]. Available: https://mentor.ieee.org/802.15/dcn/09/15-09-0276-00-0006-csem-fm-uwb-proposal.pdf

[28] Y. Zhao, Y. Dong, J.F.M. Gerrits, G. van Veenendaal, J.R. Long and J.R. Farserotu, "A Short Range, Low Data Rate, 7.2G–7.7 GHz FM-UWB Receiver Front-End," *IEEE Journal of Solid-State Circuits*, vol. 44, no. 7, pp. 1872–1882, July 2009.

[29] J.F.M. Gerrits, H. Bonakdar, M. Detratti, E. Pérez, M. Lobeira, Y. Zhao, Y. Dong, G. van Veenendaal, J.R. Long, J. R. Farserotu, E. Leroux and C. Hennemann, "A 7.2–7.7 GHz FM-UWB Transceiver Prototype," *Proc. ICUWB2009*, pp. 580–585.

[30] IST-507102 project: Magnet Beyond – My personal Adaptive Global NET, http://www.magnet.aau.dk/

2

Principles and System Design
of the Frequency Modulated UWB Radio

This chapter presents the principles of the frequency modulated ultra-wideband (FM-UWB) communication system. Section 2.2 addresses transmitter system design. It investigates the influence of subcarrier waveform, frequency and modulation scheme upon the power spectral density of the FM-UWB signal, and aims at defining optimum values for these parameters.

Section 2.3 addresses receiver system design. It mainly focuses on selection of the wideband FM demodulator and its properties. Section 2.4 closes the chapter with conclusions.

2.1 Introduction to Frequency Modulated Ultra-Wideband

Figure 2.1 shows the UWB spectral emission mask for Europe [1]. Consider the design of a UWB system to operate in the 4.2–4.8 GHz band where it is required to generate a signal with the flattest possible power spectral density and steep spectral roll-off. Steep spectral roll-off improves the coexistence of the UWB radio with other RF systems operating in adjacent frequency bands (e.g., WLAN devices operating between 5 and 6 GHz).

Frequency modulation (FM) has the unique property that the RF bandwidth (B_{RF}) is not only related to the bandwidth (f_m) of the modulating signal, but also to the modulation index (β) that can be chosen freely. This yields either a bandwidth efficient narrowband FM signal (β smaller than 1) or a (ultra) wideband signal (β much larger than 1) that can occupy any required bandwidth compatible with the RF oscillator's tuning range. The author has always been intrigued by the steep spectral roll-off of FM signals as well as the low-complexity realizations of FM radio systems, and was convinced that wideband FM could be exploited in a low-complexity UWB communication system.

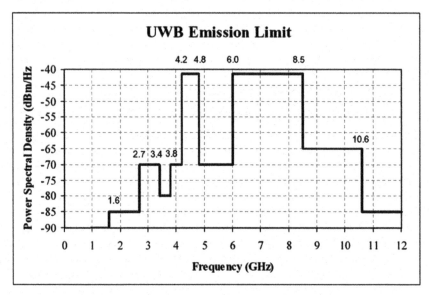

Figure 2.1   European UWB spectral emission mask [1].

In the FM-UWB approach, analog wideband FM is used as a spreading mechanism to generate a constant-envelope UWB signal of appropriate bandwidth. Digital modulation of the modulating signal m(t) by transmit data d(t) turns it into a subcarrier for data transmission. Figure 2.2 presents the block diagram of the FM-UWB transmitter. The principles of this double-FM approach, i.e., digital narrowband FSK followed by analog wideband FM, were first addressed in [2].

The wideband FM signal V(t) with amplitude A and carrier frequency f_c ($\omega_c = 2\pi f_c$) is obtained by modulating the RF oscillator with the FSK subcarrier m(t) of frequency $f_m(\omega_m = 2\pi f_m)$ such that

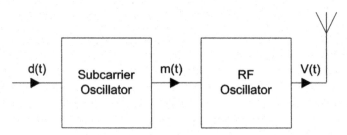

Figure 2.2   FM-UWB transmitter block diagram.

$$m(t) = V_m M \left(\omega_m t + \Delta\omega_{SUB} \int_{-\infty}^{t} d\left(\tau\right) d\tau \right). \tag{2.1}$$

M represents the triangular waveform of the subcarrier signal and $\Delta\omega_{SUB}$ is the subcarrier deviation. An RF oscillator sensitivity of K_O [rad/Vs] yields a deviation $\Delta\omega = 2\pi\Delta f$ equal to

$$\Delta\omega = K_O V_m, \tag{2.2}$$

resulting in a FM signal V(t)

$$V(t) = A \sin\left(\omega_c t + \varphi\left(t\right)\right)$$

$$= A \sin\left(\omega_c t + K_O \int_{-\infty}^{t} m\left(\tau\right) d\tau \right) \tag{2.3}$$

$$= A \sin\left(\omega_c t + \Delta\omega \int_{-\infty}^{t} M \left(\omega_m t + \Delta\omega_{SUB} \int_{-\infty}^{t} d\left(\tau\right) d\tau \right) d\tau \right)$$

Figure 2.3 shows data (d), subcarrier (m) and UWB signal (V) in the time domain for a data transition at t = 0 and subcarrier frequency of 1 MHz; the

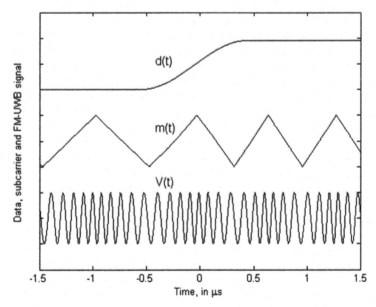

Figure 2.3 Time domain view of data, d(t), subcarrier, m(t), and UWB signal, V(t).

center frequency of the UWB signal was chosen to be 10 MHz for the sake of clarity. Binary FSK is used as the subcarrier modulation scheme.

The FM-UWB receiver block diagram is shown in Figure 2.4. The receiver demodulates the FM-UWB signal without frequency translation. The receiver comprises a LNA and a wideband FM demodulator that performs analog despreading, followed by one or several low-frequency subcarrier filtering and amplification stages and subcarrier demodulators. Due to instantaneous despreading in the wideband FM demodulator, the system behaves like a narrowband FSK system where synchronization is limited only by the bit synchronization time. As a result, fast synchronization (as required in ad-hoc networks) is obtained.

Multiple users may be accommodated in a number of ways. Apart from standard TDMA and RF FDMA techniques, FM-UWB may also use subcarrier FDMA techniques by assigning different subcarrier frequencies to different users. By avoiding hardlimiting in the receiver, simultaneous demodulation of multiple FM-UWB input signals with different subcarrier frequencies occupying the same RF band is possible. The number of simultaneous users is limited by the data rate and multiple access interference. As an illustration, six users operating at 62.5 kbps can be accommodated using FSK subcarriers between 1 and 2 MHz.

The following subsections show more details of the system design for both the transmitter and receiver, i.e., how to select system parameters to optimize radio performance and also be compliant with regulations. Section 2.2 investigates the influence of the subcarrier parameters and 2.3 presents the wideband FM demodulator for the receiver and examines the optimum demodulator bandwidth as well as the effect of frequency offset on the demodulator performance. The theoretical noise factor of the FM demodulator is also considered.

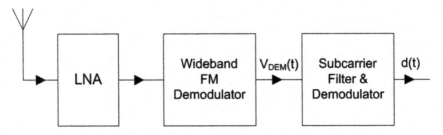

Figure 2.4   FM-UWB receiver block diagram.

2.2 FM-UWB Transmitter System Design

This section investigates the influence of subcarrier waveform, frequency and modulation scheme upon the power spectral density of the FM-UWB signals, and aims at defining optimum values for these parameters.

2.2.1 Subcarrier Waveform

The power spectral density (PSD) of a wideband FM signal is determined by and has the shape of the probability density function of the modulating signal, m(t) [3]. Modulation with a sinusoidal subcarrier signal therefore doesn't yield the best possible flatness for the power spectral density.

Triangular and sawtooth waveforms both result in a uniform probability density function and therefore yield the flattest possible RF spectrum. Is there a reason to prefer either of these two waveforms?

For any periodic subcarrier signal m(t) with fundamental period $T = 1/f_m$, the FM signal V(t) of (2.3) can be written as

$$V(t) = A \sin\left(\omega_c t + \varphi\left(t\right)\right) = Re\left(A \sum_{n=-\infty}^{\infty} c_n e^{j(\omega_c + n\omega_m)t}\right). \tag{2.4}$$

The coefficients c_n are real and given by [4]

$$c_n = \frac{1}{T} \int_0^T e^{j\phi(t)} e^{-jn\omega_m t} dt. \tag{2.5}$$

Depending on the nature of $\varphi(t)$, this integral can be evaluated either analytically or numerically. Figure 2.5 shows results of numerical calculation of the coefficients for sinusoidal, triangular and sawtooth modulating waveforms. The figure shows the absolute value of the coefficients (c_{ndB}) on a logarithmic scale (in dB), where

$$c_{ndB} = 20 \log_{10} c_n \tag{2.6}$$

The modulation index (β) is 250 and the modulation waveform is sinusoidal, triangular or sawtooth. Coefficients are shown for n ranging from –300 to 300. For a subcarrier frequency of 1 MHz, this corresponds to a 600 MHz frequency range.

As predicted by theory, the sinusoidal subcarrier doesn't yield a flat PSD. A difference of 6 dB is observed between the PSD at the center frequency and the band edges. This implies a 6 dB back-off in order to comply with the emission spectral mask.

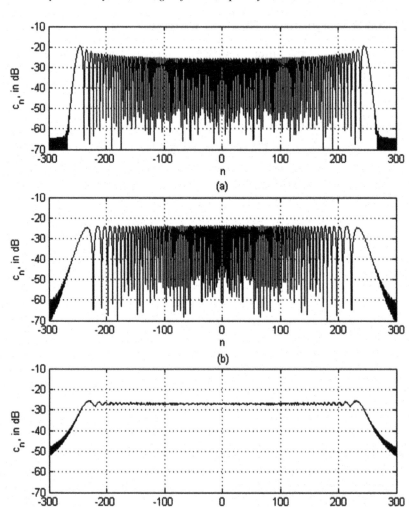

Figure 2.5 Calculated spectral coefficients of wideband FM signal modulated by (a) sinusoidal, (b) triangular and (c) sawtooth signals.

The triangular and sawtooth subcarriers both yield a flat PSD. The FM-UWB signal originating from the sawtooth has slower spectral roll-off due to the stronger harmonic contents of this waveform. As illustrated by Figure 2.5, the difference in PSD amounts to 15 dB at an offset of 300 MHz from the center frequency for a FM-UWB signal with 250 MHz deviation and

1 MHz subcarrier frequency. The steeper spectral roll-off is the first argument in favor of a triangular subcarrier waveform.

In the receiver, the FM-UWB signal is demodulated in a wideband FM demodulator. After the wideband FM demodulator the fundamental of the wanted subcarrier signal is filtered out, amplified and demodulated. By comparing the fundamental content (frequency f_m) of the demodulator output signal for the various waveforms, the subcarrier waveform yielding the highest output can be determined. Figure 2.6 shows the FM demodulator output signals for sinusoidal, triangular and sawtooth subcarrier. The triangular subcarrier waveform yields a sine wave at the FM demodulator output.

Table 2.1 presents values of the fundamental component HD_1 of these waveforms. The sinusoidal subcarrier yields the highest value for HD_1. However, the 1.1 dB advantage over the triangular waveform, corresponding

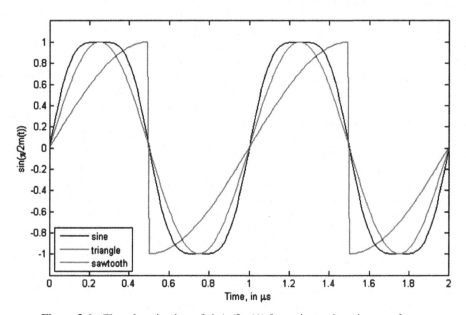

Figure 2.6 Time domain view of $\sin(\pi/2m(t))$ for various subcarrier waveforms.

Table 2.1 Fundamental content HD_1 of demodulated FM-UWB signal for various subcarrier waveforms

Waveform	HD_1, in dBV
Sinusoidal	1.10
Triangular	0.00
Sawtooth	−1.41

to a 0.55 dB sensitivity increase at RF, doesn't compensate for the 6 dB back-off required at the transmitter. Although close in performance, the triangle has a 1.4 dB advantage over the sawtooth. This gives a second argument in favor of the triangular subcarrier waveform. An additional advantage of the triangle over the sawtooth is that it lacks the abrupt transition of the sawtooth. In a practical circuit realization, such a transition may cause crosstalk and power supply noise.

It will be shown in Chapter 5 that the use of triangular local oscillator signals in the subcarrier processing part of the receiver allows for attenuation of subcarrier harmonics present in the FM demodulator output signal. This is the main motivation to choose a triangle as subcarrier signal waveform in the FM-UWB radio.

2.2.2 Subcarrier Frequency

The choice of the subcarrier frequencies (f_{SUB}) and modulation index (Δf_{SUB}) is determined by the data rate and the number of users in the FM-UWB communication system.

The subcarrier frequency needs to be higher than the data rate [4], and in order to accommodate a maximum number of users in the subcarrier FDMA multiple-access scheme, the highest possible subcarrier frequency should be used. An upper limit is inspired by regulations, as power spectral density is not allowed to exceed the −41.3 dBm/MHz limit. This value is measured with a spectrum analyzer with its resolution bandwidth set to 1 MHz, which does not permit one to distinguish individual spectral components spaced less than 1.5 MHz apart.

Increasing the subcarrier frequency at constant RF bandwidth implies lowering the modulation index (β) of the wideband FM signal. The spectrum of a FM-UWB signal of power P_{TX} and deviation Δf, is constituted by roughly 2β components of equal power $P_{TX}/2\beta$, and spaced $\Delta f/\beta$ apart. As long as $f_{SUB} < 1$ MHz, the observed PSD on the spectrum analyzer will be flat and continuous, since the number of spectral components of the FM-UWB signal falling within the 1 MHz resolution bandwidth of the analyzer is at least one.

Use of subcarrier frequencies $f_{SUB} \gg 1$ MHz will yield a spectrum analyzer display that is no longer continuous, since the spacing $\Delta f/\beta$ is now larger than the resolution bandwidth. The observed PSD is shown in Figure 2.7. In order to stay compliant with regulations, the RF power of the FM-UWB signal needs to be reduced for subcarrier frequencies above 2 MHz.

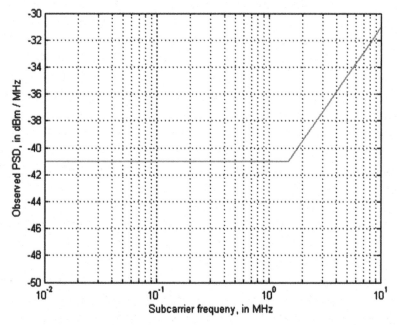

Figure 2.7   Observed power spectral density on a spectrum analyzer with 1 MHz resolution bandwidth as a function of subcarrier frequency. B&W smooth.

2.2.3 Subcarrier Modulation Scheme

It can be shown that the use of a digital, constant-envelope subcarrier modulation, typically binary or quaternary FSK, preserves the spectral flatness of the FM-UWB signal. The probability density function (PDF) of a triangular FSK signal remains uniform, yielding a flat PSD. The use of pulse amplitude modulation (PAM) or phase modulation (PM) for the subcarrier modulation scheme does not yield a uniform PDF resulting in a power density of the RF signal that is far from flat. This is because the PDF of an envelope modulated triangular wave is no longer uniform. Examples showing the PSD of the FM-UWB signal for FSK and BPSK subcarrier modulation can be found in Appendix A.

2.3 FM-UWB Receiver System Design

Since the FM-UWB signal is generated by double FM, i.e., digital FSK followed by analog wideband FM, the FM-UWB receiver needs to implement wideband FM demodulation to recover the low-frequency subcarrier signals (see Figure 2.4). In order to avoid the FM capture effect, the received RF signal is not hardlimited prior to demodulation. The capture effect causes suppression of multiple FM signals, which is undesirable in a multi-user subcarrier FDMA scenario [5]. By avoiding hardlimiting, simultaneous demodulation of multiple FM-UWB input signals with different subcarrier frequencies occupying the same RF band is possible. Performance will be limited by multiple-access interference.

2.3.1 Wideband FM Demodulator

The FM-UWB demodulator circuit should be able to process multiple signals simultaneously, where each FM-UWB signal may have a negative SNR. Not all FM demodulators operate effectively under these conditions. For example, a phase-locked loop demodulator will not capture a signal reliably when the input SNR is less than 0 dB, nor can it demodulate multiple RF signals.

The wideband FM demodulator is implemented as a fixed time delay demodulator as shown in Figure 2.8. The operation of the delay line demodulator is that of FM-to-PM conversion in the delay line followed by a multiplier operating as phase detector. This demodulator is capable of simultaneous demodulation of multiple FM signals and yields superior performance [6].

The multiplier output signal V_{DEM} equals

$$V_{DEM}(t) = A^2 \sin\left(\omega_c t + \phi(t)\right) \sin\left(\omega_c (t - \tau) + \phi(t - \tau)\right). \qquad (2.7)$$

If we ignore the high frequency term at $2\omega_c$ – a component that can be easily filtered out in a practical circuit realization – the lowpass filtered output signal V_{DEM} can be written as

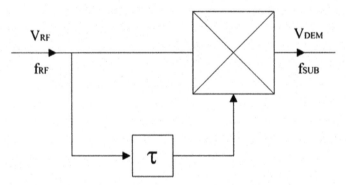

Figure 2.8   Fixed time delay FM demodulator.

$$V_{DEM}(t) = \frac{A^2}{2}\cos(\omega_c \tau + \phi(t) - \phi(t - \tau)). \qquad (2.8)$$

Choosing the delay time τ equal to an odd multiple of a quarter period (T) for the carrier frequency f_c of the FM signal

$$\tau = N\frac{T}{4} = N\frac{\pi}{2\omega_c}, \text{ with } N = 1, 3, 5 \dots \qquad (2.9)$$

and substituting this value for τ in (2.8) gives

$$V_{DEM}(t) = (-1)^{\left(\frac{N+1}{2}\right)} \frac{A^2}{2}\sin(\phi(t) - \phi(t - \tau)). \qquad (2.10)$$

When the delay τ is much smaller than the period T_m of the modulating waveform (i.e., requiring that $f_m = 1/T_m \ll f_c$), this can be rewritten as

$$V_{DEM}(t) = (-1)^{\frac{N+1}{2}} \frac{A^2}{2}\sin\left(\tau\frac{\partial\phi(t)}{\partial t}\right). \qquad (2.11)$$

Substituting (2.3) and (2.9) in (2.11) yields

$$V_{DEM}(t) = (-1)^{\left(\frac{N+1}{2}\right)} \frac{A^2}{2}\sin\left(\frac{N\pi}{2\omega_c}\Delta\omega m(t)\right). \qquad (2.12)$$

The demodulated signal is proportional to the amplitude squared of the FM signal multiplied by the sine of a constant times the original modulating signal m(t). The demodulator output voltage as a function of the input frequency, $V_{FMDEM}(f)$, is given by

$$V_{FMDEM}(f) = \frac{A^2}{2}\cos\left(\frac{N\pi f}{2f_C}\right) = -\frac{A^2}{2}\sin\left(N\frac{\pi}{2}\left[\frac{f}{f_C} - 1\right]\right). \qquad (2.13)$$

Figure 2.9 illustrates the demodulator output voltage versus the normalized input frequency (f/f_C) for various values of the parameter N. Input signal amplitude (A) is assumed to be $\sqrt{2}$.

As shown by (2.12), the demodulator sensitivity is proportional to the delay time τ. The useful RF bandwidth (B_{DEM}) of the FM demodulator, defined as the maximum frequency range over which the static demodulator transfer function is monotonic, is inversely proportional to the delay time and given by

$$B_{DEM} = \frac{1}{2\tau} = \frac{2}{N}f_c. \qquad (2.14)$$

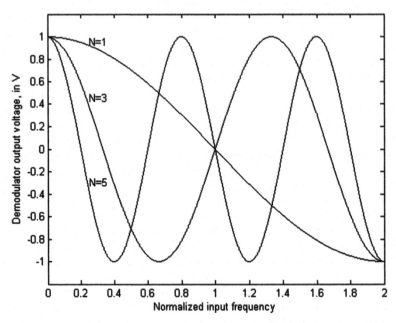

Figure 2.9   Demodulator output voltage versus normalized input frequency for various values of N.

The FM demodulator overdrive O is defined as

$$O = \frac{2\Delta f}{B_{DEM}} = N\frac{\Delta f}{f_c}. \tag{2.15}$$

(2.12) can now be rewritten as

$$V_{DEM}(t) = (-1)^{\left(\frac{N+1}{2}\right)} \frac{A^2}{2} \sin\left(O\frac{\pi}{2}M(\omega_m t)\right). \tag{2.16}$$

An overdrive of 1 corresponds to a deviation of the FM input signal equal to one-half of the FM demodulator bandwidth. With a triangular subcarrier waveform, this yields a sinusoidal demodulator output signal. Figure 2.10 shows how the fundamental (of frequency f_m) of the subcarrier in the FM demodulator output signal changes versus the overdrive for a demodulator transfer function as defined by (2.16).

For small values of the overdrive, the term $\sin\left(O\frac{\pi}{2}M(\omega_m t)\right)$ can be approximated as $O\frac{\pi}{2}M(\omega_m t)$. It can be seen that an overdrive value slightly larger than one can be tolerated since it doesn't degrade the demodulator sensitivity. In a practical FM demodulator implementation, where the delay

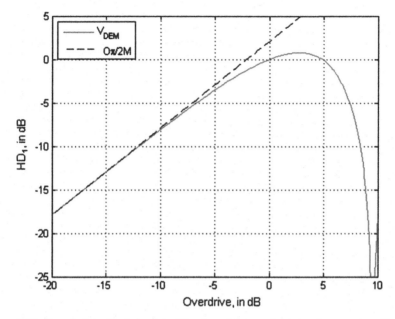

Figure 2.10 Level of the fundamental of the demodulator output signal versus overdrive for an FM demodulator with perfect sine wave transfer.

is not realized by a true delay line, but rather by a delay circuit with finite bandwidth, the sinusoidal transfer can be achieved over a limited bandwidth only, typically the bandwidth of the FM-UWB signal. In that case, an overdrive equal to unity is optimum.

2.3.1.1 Implications of Frequency Offset on the FM Demodulator Performance

In a practical radio implementation, the transmitter carrier frequency (f_{CTX}) will have an offset with respect to the demodulator center frequency (f_C) resulting in even-order harmonic distortion. The relative frequency offset (o) is defined as:

$$o = \frac{2\left(f_{CTX} - f_C\right)}{B_{DEM}}. \tag{2.17}$$

Figure 2.11 shows how DC offset HD_0, fundamental HD_1, second harmonic HD_2 and 4th harmonic HD_4 of the FM demodulator output voltage change as a function of frequency offset. A 2 dB reduction for the fundamental, equivalent to 1 dB sensitivity reduction at RF, occurs for a relative frequency offset of ± 0.4, which corresponds to ± 100 MHz offset for a demodulator bandwidth

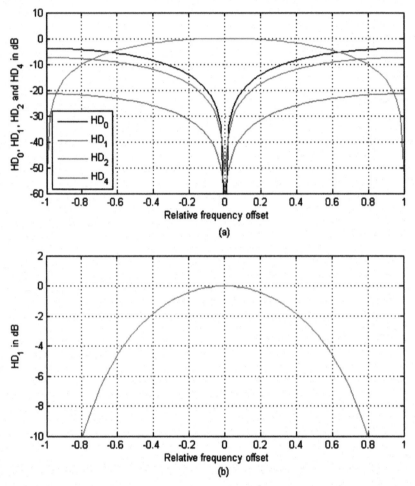

Figure 2.11   (a) DC offset HD_0, level of fundamental HD_1 second harmonic HD_2 and 4th harmonic HD_4 of the FM demodulator output voltage as a function of the relative frequency offset, (b) level of fundamental HD_1.

of 500 MHz. This implies that FM-UWB approach is tolerant to frequency offset between transmitter and receiver.

Figure 2.11 also shows that the levels of the even harmonics increase with offset. As will be shown in section 3.3, multipath is a major source of harmonic distortion at the FM demodulator output in a practical FM-UWB communication system and the harmonics caused by frequency offset can be ignored.

2.3.1.2 FM Demodulator Noise

The noise factor (F) of an amplifier or mixer is defined as the degeneration of the signal-to-noise ratio (SNR) from the input to output of the device.

$$F = \frac{SNR_I}{SNR_O} \tag{2.18}$$

Input and output SNR are measured in the same bandwidth (e.g., 1 Hz), but not necessarily at the same frequency. For example, when the circuit performs a frequency translation (up or downconversion), the input and output SNR are measured at different frequencies.

Usually noise factor and noise figure (the noise factor expressed in dB), are defined for linear systems where the output signal is proportional to the input signal and noise figure is independent of signal level. Output SNR (in dB) equals input SNR (in dB) minus the noise figure.

However, the wideband FM demodulator based upon the delay line demodulator (see Figure 2.8) is a non-linear, quadratic device. Therefore, the conventional definition of noise factor may not seem to be directly applicable here. Section 3.1 presents the exact relation between input (SNR_{RF}) and output SNR (SNR_{DEM}) of the wideband demodulator. The noise factor of the demodulator is found to be

$$F = \frac{SNR_{RF}}{SNR_{DEM}} = 4 + \frac{1}{SNR_{RF}}. \tag{2.19}$$

The noise factor is dependent on the input SNR as shown in Figure 2.12 and it converges to a constant value of 4 (NF = 6 dB) for higher SNR values. All the above assume a *noiseless* FM demodulator. In reality such a demodulator does not exist. Analysis of the noise figure of realistic wideband FM demodulator circuits, using, for example, a double-balanced Gilbert mixer as multiplier, shows that their noise contribution can be relatively high. More details on the noise of practical multiplier circuits will be presented in Chapter 5.

2.4 Conclusions

This chapter has presented the principles of the frequency modulation ultra-wideband (FM-UWB) communication system. It was shown that the triangular subcarrier waveform has advantages for the power spectral density and roll-off of the FM-UWB signal. Constant-envelope subcarrier modulation schemes, like 2-FSK or 4-FSK are required to obtain a flat power spectral

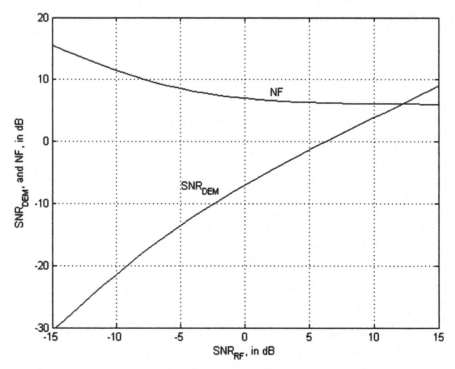

Figure 2.12 SNR of FM demodulator output signal and noise figure of noiseless FM demodulator.

density. The optimum subcarrier frequency lies between 1 and 2 MHz. Higher subcarrier frequencies will require back-off in order to comply with regulations.

The fixed time delay FM demodulator is the best candidate for simultaneous demodulation of multiple FM-UWB signals with possibly negative SNR. It is best used with an overdrive equal to unity, i.e., the demodulator bandwidth equals the FM-UWB signal bandwidth. The FM-UWB approach is tolerant to frequency offset between transmitter and receiver. A 1 dB sensitivity reduction at RF occurs for ±100 MHz offset for a demodulator bandwidth of 500 MHz. The noise factor of a noiseless FM demodulator is dependent on the input SNR as shown in Figure 2.12 and it converges to a constant value of 4 (NF = 6 dB) for higher SNR values.

Due to instantaneous despreading in the wideband FM demodulator, the FM-UWB radio behaves like a narrowband FSK system where synchronization is limited only by the receiver bit synchronization time.

References

[1] ECC Electronic Communications Committee, *ECC Decision of 24 March 2006 on the harmonised conditions for devices using Ultra-Wideband (UWB) technology in bands below 10.6 GHz*, ECC/DEC/(06)04, APT/AWF (Amended ECC/DEC/(06)04), [Online]. Available: http://www.apwpt.org/downloads/eccdec0604uwb.pdf

[2] John F.M. Gerrits, Michiel H.L. Kouwenhoven, Paul R. van der Meer, John R. Farserotu, John R. Long, "Principles and Limitations of Ultra Wideband FM Communications Systems," *EURASIP Journal on Applied Signal Processing*, vol. 2005, no. 3, pp. 382–396, March 2005.

[3] Nelson M. Blachman, George A. McAlpine, "The Spectrum of a High-Index FM Waveform: Woodward's Theorem Revisited," *IEEE Transactions on Communication Technology*, vol. 17, no. 2, pp. 201–208, April 1969.

[4] Ralph S. Carson, *Radio Communications Concepts: Analog*, New York: John Wiley & Sons, 1990.

[5] K. Leentvaar and J.H. Flint, "The Capture Effect in FM Receivers," *IEEE Transactions on Communications*, vol. 24, no. 5, pp. 531–539, May 1976.

[6] M.H.L. Kouwenhoven, *High-Performance Frequency-Demodulation Systems*, PhD thesis, Technische Universiteit Delft, 1998.

3

Performance of Frequency Modulated UWB

This chapter presents the theoretical performance of the FM-UWB communication system under various conditions. Section 3.1 discusses performance with AWGN. Section 3.2 investigates multi-user FM-UWB systems. Section 3.3 investigates the effect of frequency-selective multipath and presents the performance with the IEEE UWB and BAN channel models. Section 3.4 addresses in-band and out-of-band interference as well as interference mitigation strategies. Section 3.5 presents the conclusions.

3.1 FM-UWB Performance with AWGN

This section illustrates how the double FM system performs under additive white Gaussian noise (AWGN) conditions. Figure 3.1 shows a receiver block diagram useful for calculating the probability of error of the digital output signal.

It consists of a cascade of wideband FM demodulator, subcarrier filter, and subcarrier demodulator. The wideband FM demodulator acts as a SNR converter. The SNR at the wideband demodulator output (SNR_{DEM}) is a nonlinear function of the input SNR (SNR_{RF})

$$SNR_{DEM} = \Psi\left(SNR_{RF}\right). \tag{3.1}$$

Next the bandwidth of the demodulated signal B_{DEM} is limited to the bandwidth B_{SUB} of the FSK subcarrier signal in the subcarrier bandpass filter. The subcarrier SNR is given by

$$SNR_{SUB} = \frac{B_{DEM}}{B_{SUB}} SNR_{DEM}. \tag{3.2}$$

The subcarrier SNR determines the probability of error at the subcarrier demodulator output. Assuming binary FSK with coherent detection for the subcarrier modulation scheme, the probability of error p_b equals [1]

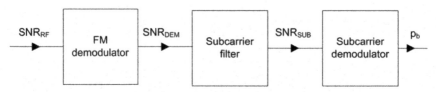

Figure 3.1   Receiver block diagram for determining the output probability of error.

$$p_b = \frac{1}{2}erfc\sqrt{\frac{SNR_{SUB}}{2}}. \tag{3.3}$$

The calculation of the SNR transfer function, Ψ, of the FM demodulator is based upon the model presented in Figure 3.2. The exact mathematical calculation is rather tedious [2]; the noise conversion depends on the frequency offset. The following intuitive calculation assumes a frequency offset equal to zero and autocorrelation of the noise, $R_N(\tau)$, equal to one for values of the delay τ according to (2.9). Although a rather coarse approximation, it yields results that correspond well to measurements made by the author on a hardware prototype of the wideband demodulator.

The FM demodulator's input voltage V_{RF} consists of the sum of signal voltage V_S and noise voltage V_N. The signal power (S) is proportional to $V_S{}^2$ and the noise power (N) is proportional to $V_N{}^2$. The bandwidth of both signals is B_{RF}.

In the intuitive calculation, the FM demodulator's output voltage V_{DEM} is approximated as

$$V_{DEM} = V_{RF}^2 = (V_S + V_N)^2$$
$$= V_S^2 + V_N^2 + 2V_S V_N$$

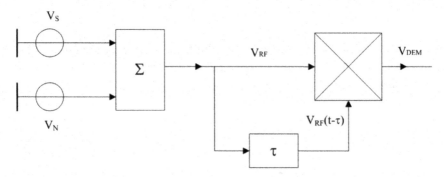

Figure 3.2   FM demodulator with signal and noise sources at its input.

$$\begin{aligned}&= S + N + 2\sqrt{S}\sqrt{N}\\&= V_{SO} + V_{N1O} + V_{N2O}\end{aligned} \tag{3.4}$$

The output voltage V_{DEM} is the sum of signal voltage V_{SO} and two independent noise voltages V_{N1O} and V_{N2O}. The power of each of these three independent terms is proportional to the square of the voltage. Output signal power (S_O) and total output noise power (N_O) are equal to

$$S_O = V_{SO}^2 = S^2, \tag{3.5}$$

$$N_O = V_{N1O}^2 + V_{N2O}^2 = N^2 + 4SN. \tag{3.6}$$

The demodulator output SNR (SNR_{DEM}) is then given by

$$SNR_{DEM} = \psi(SNR_{RF}) = \frac{S_O}{N_O} = \frac{S^2}{N^2 + 4SN} = \frac{SNR_{RF}^2}{1 + 4SNR_{RF}}. \tag{3.7}$$

This expression can be approximated as

$$SNR_{DEM} \approx SNR_{RF}^2 \text{ for } SNR_{RF} << \frac{1}{4}, \tag{3.8}$$

$$SNR_{DEM} \approx \frac{1}{4}SNR_{RF} \text{ for } SNR_{RF} >> \frac{1}{4}. \tag{3.9}$$

(3.8) corresponds to operation "below threshold" where the output SNR is proportional to the square of the input SNR [3]. (3.9) corresponds to operation "above threshold" where the output SNR is proportional to the input SNR. Assuming a flat spectrum for the noise terms V_{N1O} (noise x noise) and V_{N2O} (noise x signal) of bandwidth B_{RF}, and referring to Figure 3.1, it follows for the subcarrier SNR (SNR_{SUB}) at the input of the FSK demodulator

$$\begin{aligned}SNR_{SUB} &= \frac{B_{RF}}{B_{SUB}}SNR_{DEM} = G_P SNR_{DEM} = G_P \Psi(SNR_{RF})\\&= G_P \frac{SNR_{RF}^2}{1 + 4SNR_{RF}}.\end{aligned} \tag{3.10}$$

The receiver processing gain, G_P, is defined as the ratio of the RF and subcarrier bandwidth

$$G_P = \frac{B_{RF}}{B_{SUB}} = \frac{2\Delta f_{RF}}{(\beta_{SUB} + 1)R}. \tag{3.11}$$

Figure 3.3 shows subcarrier SNR versus RF SNR for data rates R from 1 to 1000 kbps. The subcarrier modulation index β_{SUB} is constant and equal to one,

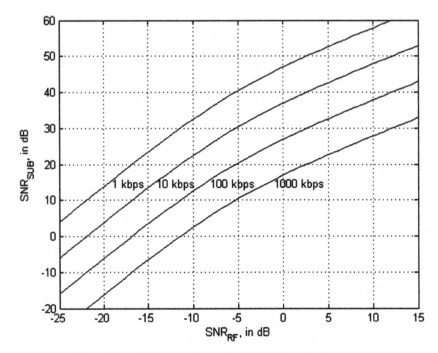

Figure 3.3   SNR conversion with AWGN at various data rates.

resulting in a subcarrier bandwidth of twice the data rate. Since the subcarrier bandwidth is proportional to the data rate, the curves in Figure 3.3 are spaced 10 dB apart.

Figure 3.4 shows the probability of error as a function of the RF SNR for the FM-UWB system with constant 500 MHz bandwidth for data rates from 1 kbps to 1000 kbps. For comparison, the figure also shows the probability of error for a narrow band binary FSK system occupying a RF bandwidth B_{SUB}. Such a signal can be obtained by replacing the RF VCO in the transmitter (see Figure 2.2) by an upconversion stage. In the narrowband FSK system case, no SNR conversion occurs and the four curves coincide.

A fair comparison between FM-UWB and FSK can be made for signals having equal signal power (resulting in equal energy per transmitted bit, E_b) and equal receiver noise single-sided power density (N_0). It can be shown that

$$\frac{E_b}{N_O} = \frac{S}{R}\frac{B_{RF}}{N} = SNR_{RF}\frac{B_{RF}}{R}. \qquad (3.12)$$

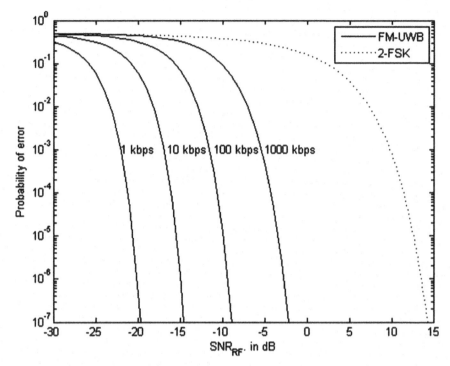

Figure 3.4 Probability of error versus RF SNR.

For a conventional modulation scheme like FSK, the RF signal bandwidth is proportional to the transmission rate, R. For the FM-UWB system, the RF bandwidth is constant.

Figure 3.5 shows results of such a comparison; the probability of error is shown versus E_b/N_0. It can be concluded that for low data rates there is a considerable penalty in receiver performance. The reason is that for low data rates the wideband FM demodulator is always operating below threshold. At medium data rates, the situation gets better and the difference between FSK and FM-UWB lowers to 13 dB at a data rate of 1000 kbps.

At even higher data rates the difference remains constant, since now the wideband FM demodulator operates above threshold and the subcarrier SNR increases linearly with the RF SNR. The performance degradation is also dependent on the subcarrier modulation index. A smaller value of β_{SUB} lowers the performance penalty. Figure 3.6 illustrates this phenomenon for a fixed data rate of 1000 kbps and subcarrier modulation index values of 0.5, 1, 2, 4 and 8.

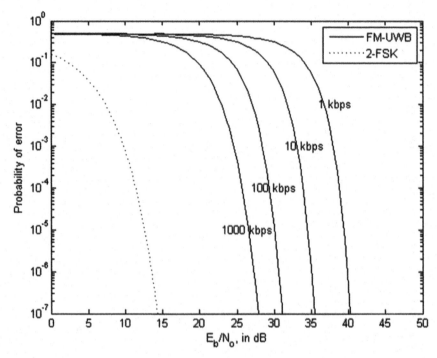

Figure 3.5   Probability of error versus E_b/N_0 of error comparison for FM-UWB and FSK.

3.1.1 Link Span

We will now examine the link budget of a typical FM-UWB communication system operating at 7.5 GHz with a bandwidth of 500 MHz, data rate R, subcarrier modulation index of one, at an error probability of 1×10^{-6}.

As a first step, the required RF SNR required to obtain a subcarrier SNR of 14 dB (yielding $p_b = 1 \times 10^{-6}$) is calculated. Substituting a subcarrier bandwidth of 2R into (3.10), and defining the inverse of the SNR conversion function Ψ as Ψ^{-1}, it follows that

$$SNR_{RF} = \Psi^{-1}\left(\frac{B_{SUB}}{B_{RF}}SNR_{SUB}\right) = \Psi^{-1}\left(\frac{2R}{B_{RF}}SNR_{SUB}\right). \quad (3.13)$$

One finds values for SNR_{RF} between –20 dB (at 1 kbps) and –3 dB (at 1000 kbps).

The second step is to find the pathloss that yields these SNR values. The received signal power at the receiver input depends on the transmit power and the pathloss (PL), defined as the ratio of transmitted power P_{TX} and received

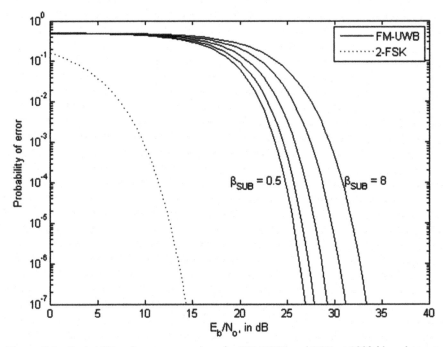

Figure 3.6   Probability of error comparison for FM-UWB and FSK at 1000 kbps data rate and subcarrier modulation index values of 0.5, 1, 2, 4 and 8.

power P_{RX}. The pathloss as a function of distance (d) between transmitter and receiver is given by the Friis transmission equation

$$PL = \frac{P_{TX}}{P_{RX}} = \frac{(4\pi d)^2}{G_{TX}G_{RX}\lambda^2}, \tag{3.14}$$

with λ the wavelength ($\lambda = 4$ cm at 7.5 GHz) and G_{TX} the transmit antenna gain and G_{RX} the receive antenna gain. Figure 3.7 shows the received power as a function of the distance under free space propagation conditions, assuming isotropic antennas ($G_{TX} = G_{RX} = 1$). Transmit power equals –14.3 dBm; the maximum power allowed for a UWB system with 500 MHz bandwidth.

The received power (P_{RX}) can be written as

$$P_{RX} = SNR_{RF}P_{NRX}. \tag{3.15}$$

The equivalent receiver input noise power P_{NRX} for an RF bandwidth B_{RF} and a receiver noise factor NF is equal to

$$P_{NRX} = kTFB_{RF}. \tag{3.16}$$

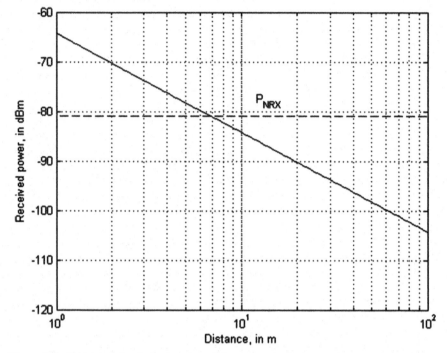

Figure 3.7   Received signal as a function of distance under free space propagation conditions at 7.5 GHz and equivalent input receiver noise power.

Assuming a receiver noise factor of 3 (5 dB noise figure), this yields an equivalent noise power P_{NRX} of –82 dBm at the receiver input. The pathloss can now be written as

$$PL = \frac{P_{TX}}{P_{RX}} = \frac{P_{TX}}{kTFB_{RF}SNR_{RF}}. \tag{3.17}$$

For a subcarrier SNR of 14 dB, and a bit rate varying between 1 and 1000 kbps, a pathloss between 88 dB and 71 dB can be dealt with, as shown in Table 3.1. Combining (3.14) and (3.17) and solving for the distance (d) yields

$$d = \frac{\lambda}{4\pi} \sqrt{\frac{P_{TX}G_{TX}G_{RX}}{kTFB_{RF}SNR_{RF}}}. \tag{3.18}$$

Transmit power P_{TX} is given by the product of the maximum allowed power spectral density (fixed by regulations at –41.3 dBm/MHz) and the RF bandwidth

Table 3.1 Required RF SNR, pathloss and equivalent free space distance to obtain $p_b = 1 \times 10^{-6}$ for data rates from 1 kbps to 1000 kbps

Data Rate, in kbps	SNR_{RF}, in dB	PL_{dB}, in dB	d_{FS}, in m
1	−20.4	87.9	79.1
10	−15.1	82.7	43.3
100	−9.5	77.0	22.5
250	−7.0	74.5	16.9
1000	−2.8	70.3	10.5

$$P_{TX} = PSD_{MAX} B_{RF}. \tag{3.19}$$

Substituting (3.13) and (3.19) into (3.18) yields

$$d = \frac{\lambda}{4\pi} \sqrt{\frac{PSD_{MAX} G_{TX} G_{RX}}{kTF\Psi^{-1}\left(\frac{2R}{B_{RF}} SNR_{SUB}\right)}} \tag{3.20}$$

The function $\Psi^{-1}(x)$ may be approximated by $\sqrt{x}$ for values of x much smaller than 1/16, which yields the following approximation for the distance

$$d \approx \frac{\lambda}{4\pi} \sqrt{\frac{PSD_{MAX} G_{TX} G_{RX}}{kTF}} \frac{1}{\sqrt[4]{2SNR_{SUB}\frac{R}{B_{RF}}}}. \tag{3.21}$$

The error of this approximation is smaller than 10 % at 100 kbps. For the lower bit rates, increasing the bit rate by a factor of 10 yields a distance reduction of $\sqrt[4]{10} = 1.78$. Table 3.1 shows how RF SNR, allowable path loss and the equivalent link span (d_{FS}) under free space propagation conditions vary with bit rate. The table shows the exact numbers and doesn't use the approximation for $Y^{-1}(x)$.

These figures indicate a good link margin for short-range LDR and MDR PAN and BAN applications. The performance of the FM-UWB system is degraded by approximately 15 dB with respect to narrowband FSK systems, however, in return the FM-UWB approach provides robustness against interference and multipath that is not available in narrowband FSK systems.

A more practical link budget for the body-area network application will be presented in Chapter 5.

3.1.2 Influence of RF Bandwidth on Link Span

How is the link span influenced by the bandwidth of the FM-UWB signal? The transmit power is proportional to the bandwidth, as in (3.19). The equivalent receiver input noise power P_{NRX} is given by (3.13) and proportional to the RF bandwidth. As a consequence, the SNR at receiver input (SNR_{RF}) is not affected by the signal bandwidth.

Assuming that the overdrive for the FM demodulator remains constant, the SNR after FM demodulation (SNR_{DEM}) remains unchanged. The processing gain as given by (3.20), is proportional to the FM-UWB signal bandwidth. Doubling the signal bandwidth yields a 3 dB subcarrier SNR increase in the receiver. This corresponds to a 1.5 dB sensitivity increase at RF. The link span d_{FS} is multiplied by a factor 1.2.

3.2 FM-UWB Performance with Multiple Users

FM-UWB exploits subcarrier FDMA where individual users distinguish themselves by different subcarrier frequencies. We will now investigate the multiple-access interference occurring in the FM-UWB system using this access scheme.

Figure 3.8 schematically shows the wideband FM demodulator with N FM-UWB signals $V_1(t)$, $V_2(t)$, to $V_N(t)$ at its input. The wideband demodulator output signal V_O comprises N^2 terms:

- N terms of the form $V_i(t)V_i(t - \tau)$ (i = 1, 2, ..., N) constituting the sum of the N subcarrier signals and
- N(N – 1) terms of the form $V_i(t)V_j(t - \tau)$ (i = 1, 2...N; j = 1, 2, ..., N, j ≠ i) constituting the multiple-access interference residue.

It is assumed that all N users are operating at the same center frequency (f_C) and use the same deviation ($\triangle f$). Each of them has its particular and unique

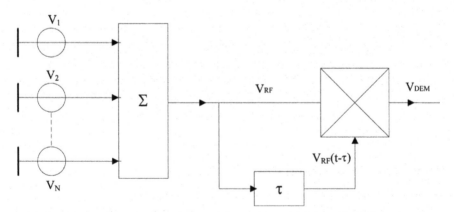

Figure 3.8 Wideband delay line demodulator with multiple FM-UWB input signals.

subcarrier frequency f_{SUBi}. Two cases are of particular interest in a multi-user environment:

1. 2 FM-UWB signals of unequal power ($S_1 \neq S_2$),
2. N FM-UWB signals of equal signal power ($S_1 = S_2 = S_3 = \ldots S_N$).

These two cases will be addressed in the following subsections.

3.2.1 Analysis of a 2-User System

An often occurring interference scenario is that of the stronger FM-UWB signal V_2 causing interference to the desired signal V_1. The following intuitive calculation provides results that correspond well with both simulation and measurement results. Referring to Figure 3.8, the FM demodulator's output voltage V_{DEM} can be approximated as

$$V_{DEM} = (V_1(t) + V_2(t))(V_1(t-\tau) + V_2(t-\tau))$$

$$\approx V_1^2(t) + V_2^2(t) + 2V_1(t)V_2(t)$$
$$= S_1 + S_2 + 2\sqrt{S_1}\sqrt{S_2} \qquad (3.22)$$
$$= V_{S1} + V_{S2} + V_{NO}$$

The output voltage V_{DEM} is the sum of two signal terms V_{S1} and V_{S2} plus a noise term V_{NO} that is a combination of FM and AM signal. The instantaneous frequency of this multiple access residue is proportional to the difference of the two subcarrier signals, $m_1(t)$ and $m_2(t)$, and its envelope is proportional to the sum of the two subcarrier signals. Figure 3.9 shows the two subcarrier signals, $m_1(t)$ and $m_2(t)$, of equal amplitude, their sum and difference and the multiple-access residue for a 2-user case with subcarrier frequencies of 1 and 1.25 MHz. The subcarrier waveform is a sine wave for the sake of clarity.

The output signal power S_O (subcarrier 1) and the output noise power N_O are equal to

$$S_O = S_1^2 \qquad (3.23)$$
$$N_O = 4S_1S_2 \qquad (3.24)$$

The SNR at the demodulator output is given by

$$SNR_{DEM} = \frac{S_O}{N_O} = \frac{S_1^2}{4S_1S_2} = \frac{S_1}{4S_2} = \frac{SIR}{4} \qquad (3.25)$$

with the signal to interference ratio defined as the ratio of the signal power of the two FM-UWB signals (i.e., $SIR = S_1/S_2$).

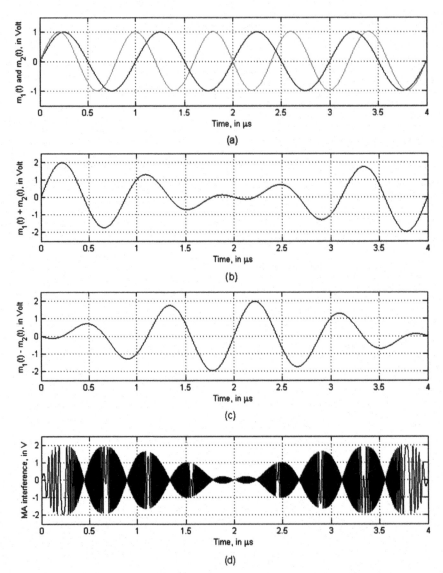

Figure 3.9   (a) Subcarrier signals $m_1(t)$ and $m_2(t)$, (b) $m_1(t) + m_2(t)$, (c) $m_1(t) - m_2(t)$, and (d) multiple access residue for $f_{SUB1} = 1$ MHz and $f_{SUB2} = 1.25$ MHz.

The residue signal V_{NO} is the result of the two FM-UWB signals down-converting each other yielding a FM signal centered at zero. Its single-sided power spectral density is flat and has a bandwidth equal to the RF bandwidth of the FM-UWB signal ($B_{RF} = 2\triangle f$). Only the part of the residue falling within the subcarrier bandwidth contributes to the degradation of the subcarrier SNR. The receiver processing gain, G_P, is defined in (3.11) as the ratio of the RF and subcarrier bandwidth.

As with the AWGN case, the subcarrier SNR is given by

$$SNR_{SUB}\frac{B_{RF}}{B_{SUB}}SNR_{DEM} = G_P\frac{SIR}{4}. \tag{3.26}$$

Figure 3.10 shows how the processing gain depends on the bit rate and the bandwidth of the FM-UWB signal. For example, in a 250 kbps FM-UWB system with a RF bandwidth of 500 MHz and a subcarrier modulation index β_{sub} of 1, the receiver processing gain equals 30 dB. In the case of a strong FM-UWB interferer, the probability of error increases to 1×10^{-3} for a subcarrier SNR of 10 dB, corresponding to an interferer 20 dB stronger than the wanted signal.

CW jammer is like MA interference, but 3 dB lower, since bandwidth of residue signal is equal to half the bandwidth of the FM-UWB signal.

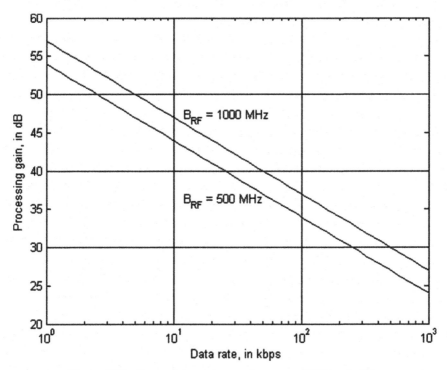

Figure 3.10 Processing gain versus data rate and RF bandwidth.

3.2.2 Analysis of a N-User System

We will now investigate the situation of N FM-UWB signals of equal signal power ($S_1 = S_2 = S_3 = \ldots S_N = 1$ where $S_i = V_i^2$). The FM demodulator output voltage V_{DEM} is approximated as

$$
\begin{aligned}
V_{DEM} = V_{RF}^2 &= \left(\sum_{i=1}^{N} V_i \right)^2 \\
&= \sum_{i=1}^{N} V_i^2 + \sum_{*} V_i V_j \\
&= V_1^2 + V_2^2 + \ldots + V_N^2 + V_1 V_2 + V_1 V_3 + \ldots + V_N V_{N-1} \\
&= \sum_{i=1}^{N} S_i + \sum_{*} \sqrt{S_i} \sqrt{S_j} \\
&= \sum_{i=1}^{N} V_{Si} + V_{NO}
\end{aligned}
\tag{3.27}
$$

where *: (i = 1, 2, ... , N; j = 1, 2, ... , N, j ≠ i).

The output voltage V_O is the sum of N signal terms $V_{S1}, V_{S2} \ldots V_{SN}$ plus the multiple-access interference term V_{NO}. The output signal power S_O (e.g., subcarrier 1) and the output noise power N_O are equal to

$$
S_O = S_1^2
\tag{3.28}
$$

$$
N_O = 2 \sum_{*} S_i S_j
\tag{3.29}
$$

When all signal powers are normalized to unity, the SNR at the demodulator output is given by

$$
SNR_{DEM} = \frac{S_O}{N_O} = \frac{1}{2N(N-1)}.
\tag{3.30}
$$

The residue signal V_{NO} is again the result of ½N(N − 1) FM-UWB signal pairs downconverting each other in the frequency domain. The power spectral density of this FM signal is flat and has a bandwidth B_{RF} equal to $2\Delta f$. Only the part of the residue falling within the subcarrier bandwidth contributes to the degradation of subcarrier SNR. It follows for the subcarrier SNR

$$
SNR_{SUB} = G_P SNR_{DEM} = \frac{G_P}{2N(N-1)},
\tag{3.31}
$$

or expressed in dB

$$SNR_{SUBdB} = G_{PdB} - 10\log_{10}\left(N\left(N-1\right)\right) - 3dB. \qquad (3.32)$$

As an illustration in a 5-user 100 kbps FM-UWB system with an RF bandwidth of 500 MHz and a subcarrier modulation index β_{sub} of 1, the subcarrier SNR degrades to 18 dB. Clearly, there are a number of users for which the multiple-access interference degrades the bit error rate (BER) of the demodulated FSK subcarrier to an unacceptable value. By choosing a probability of error limit (e.g., at 1×10^{-3}) which corresponds to a minimum SNR of 10 (10 dB), it is possible to calculate the number of users of equal power that will yield this value by solving (3.29) for N equal to N_{MAX} and SNR_{SUB} equal to SNR_{SUBMIN}, or

$$2N_{MAX}\left(N_{MAX} - 1\right) = \frac{G_P}{SNR_{SUBMIN}}. \qquad (3.33)$$

Approximating $N(N-1)$ by N^2 yields the following approximation for the maximum number of users N_{MAXMA} in a FM-UWB system that is multiple-access interference limited.

$$N_{MAXMA} \approx \sqrt{\frac{G_P}{2SNR_{SUBMIN}}} = \sqrt{\frac{1}{2SNR_{SUBMIN}}\frac{B_{RF}}{(\beta_{SUB}+1)\,R}}. \qquad (3.34)$$

It is important to note that to actually accommodate N_{MAXMA} users at bit rate R, a minimum subcarrier frequency f_{SUBMIN} is required where

$$f_{SUBMIN} = N_{MAX}R(\beta_{SUB}+1). \qquad (3.35)$$

As shown in Section 2.2.2, the use of subcarrier frequencies above 2 MHz requires reduction of transmit power, which lowers the link span. A trade-off needs to be made and the result may be that the subcarrier frequency range is restrained. The number of users will be limited to a lower value N_{MAXSUB}, equal to

$$N_{MAXSUB} = \frac{f_{SUBMIN}}{R(\beta_{SUB}+1)}. \qquad (3.36)$$

Consequently, the maximum number of users in a FM-UWB communication system based upon subcarrier FDMA techniques is given by the minimum of N_{MAXMA} and N_{MAXSUB}

$$N_{MAX} \approx \min\left(N_{MAXMA}, N_{MAXSUB}\right)$$

$$= \min\left(\sqrt{\frac{1}{2SNR_{SUBMIN}}\frac{B_{RF}}{(\beta_{SUB}+1)R}}, \frac{f_{SUBMIN}}{R(\beta_{SUB}+1)}\right).$$

$$(3.37)$$

Figure 3.11 illustrates this equation for a RF bandwidth of 500 MHz, SNR_{SUBMIN} = 10 dB, β_{SUB} = 1 and subcarrier frequency range of 1–2 MHz. Up to 10 kbps, the number of users is limited by multiple-access interference and for higher data rates by the available subcarrier bandwidth. The solid curve represents N_{MAX}.

3.2.3 FM-UWB Capacity Analysis

The capacity (C_{MAX}) of the FM-UWB communication system is obtained by multiplying the maximum number of users by the data rate yielding

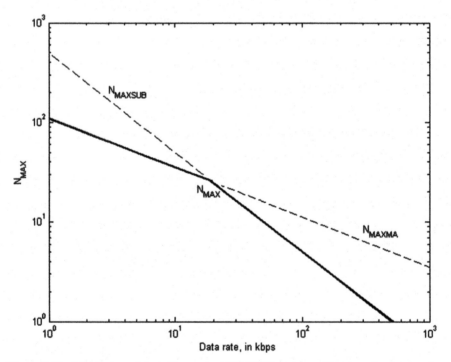

Figure 3.11   Maximum number of users in a FM-UWB system with subcarrier frequencies between 1 and 2 MHz.

$$C_{MAX} = N_{MAX} R. \tag{3.38}$$

Capacity, like the maximum number of users, is either limited by the multiple access interference or the subcarrier frequency range. The total capacity C_{MAX} is given by (3.37) multiplied by R, or

$$
\begin{aligned}
C_{MAX} &= \min\left(C_{MAXMA}, C_{MAXSUBBW}\right) \\
&\approx \min\left(\sqrt{\frac{R}{2SNR_{SUBMIN}}} \frac{B_{RF}}{(\beta_{SUB}+1)}, \frac{f_{SUBMIN}}{(\beta_{SUB}+1)}\right).
\end{aligned}
\tag{3.39}
$$

Figure 3.12 shows the capacity for a system using a subcarrier range of 1–2 MHz and a 500 MHz RF bandwidth. A 100 kbps FM-UWB system with its subcarrier frequencies between 1 and 2 MHz and subcarrier modulation index of 1, can accommodate 5 users. If we lower the bit rate to 10 kbps, the number of users increases to 50. The capacity is limited to an aggregate total of 500 kbps.

Higher capacity values can be obtained by increasing the subcarrier frequency or by using higher order FSK modulation. Unfortunately, as shown

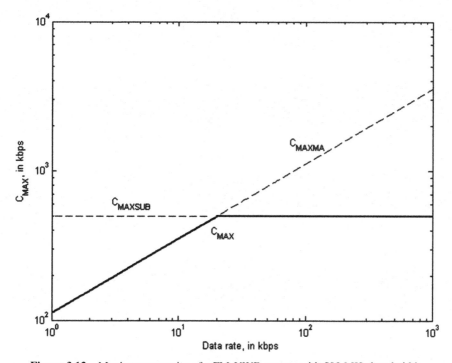

Figure 3.12   Maximum capacity of a FM-UWB system with 500 MHz bandwidth.

in Section 2.2.2, increasing the subcarrier frequency above 2 MHz, will require reduced transmission power to stay compliant with regulations. Lowering the modulation subcarrier modulation index may also seem to be an option. However, subcarrier modulation index values lower than 1 will lower the performance of the FSK scheme. Therefore the practical capacity limit for a low-complexity FM-UWB system using binary FSK as subcarrier modulation scheme is $\frac{1}{2}f_{SUBMIN}$.

3.2.4 Performance Limitations in a Subcarrier FDMA System

When using a subcarrier FDMA scheme, there are four reasons why a wanted signal may be drowned by a stronger (unwanted) signal using the same RF center frequency and a different subcarrier frequency: multiple-access interference, insufficient sidelobe rejection in the transmitter subcarrier generation, transmitter RF oscillator phase noise, and insufficient selectivity of the receiver subcarrier filtering.

The multiple-access interference is a fundamental limitation, whereas the other three depend on circuit implementation.

3.3 FM-UWB Performance with Frequency-Selective Multipath

Figure 3.13 presents a simplified model of the RF propagation from transmitter to receiver [4]. It is a cascade of a frequency independent part representing the path loss (PL) and a frequency-dependent channel transfer function H(f) that also includes the transfer function of the antennas, and any RF filtering prior to the FM demodulator. For free space propagation conditions, wideband constant-aperture antennas and no in-band filtering, the channel transfer function, H(f) is approximately one.

In the case of multipath, the channel transfer function, H(f), will not be flat. The UWB signal is scanning this frequency dependent channel transfer

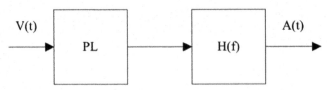

Figure 3.13 Propagation model with frequency independent part PL and frequency-dependent part H(f).

function at the rhythm of the subcarrier signal m(t). The amplitude A(t) of the received UWB signal will therefore vary accordingly. As a result of these amplitude variations, the output signal at the wideband FM demodulator will show harmonic distortion, and the signal power of the fundamental (HD_1) of the subcarrier will depend on the channel transfer function. The subcarrier SNR will also vary accordingly. Noise in the subcarrier bandwidth doesn't depend on the channel since it is the result of thermal noise at the receiver input and the noise of the receiver front-end.

The fading margin is defined by the increase in required signal power to realize the same probability of error on fading versus non-fading channels [5]. The fading margin is given at a defined availability, typically 99% of the time.

Radio performance in a specific multipath environment can be benchmarked in simulation using channel models [6]. Channel models constitute a statistical model of the propagation environment and are derived from measurements made in one or several specific propagation scenarios, e.g., line-of-sight, non line-of-sight, body-worn antennas, etc. The model comprises an algorithm to generate a set of time-domain impulse responses based upon parameters corresponding to specific propagation scenarios. A variety of channel models for UWB applications are available from the various IEEE task groups (e.g., IEEE802.15.3a [7], IEEE802.15.4a [8], IEEE802.15.6 [9]).

The next subsections present the effect of frequency selective multipath on the performance of the FM-UWB communication system, and a methodology to efficiently simulate the performance of the FM-UWB system for any propagation channel. Simulation results are presented for the IEEE802.15.3a indoor channel models and IEEE802.15.6 BAN channel models. Finally, an intuitive explanation for good and bad propagation channels will be elaborated, and an attempt is made to identify the best and worst possible channels.

3.3.1 Consequences of Frequency-Selective Multipath

As already mentioned in Section 2.3, the demodulator output voltage for an input signal of amplitude A and frequency f modulated by a subcarrier signal m(t) of unity amplitude is given by

$$V_{DEM}(t) = K_D A^2 \sin\left(O\frac{\pi}{2}m\left(t\right)\right), \tag{3.40}$$

where K_D is the demodulator sensitivity. The instantaneous amplitude, A(t), of the receiver input signal is equal to the channel transfer function value of the instantaneous frequency (f_{INST}) of the FM-UWB signal

$$A(t) = H(f_{INST}(t)) = H(f_C + \Delta f m(t)). \tag{3.41}$$

With an overdrive of 1 (i.e., $B_{RF} = B_{DEM}$, meaning optimum use of the demodulator), (3.38) can be rewritten as

$$V_{DEM}(t) = K_D A^2(t) \sin(m(t)) = K_D B(t) \sin(m(t)). \tag{3.42}$$

This is illustrated in Figure 3.14 for a subcarrier frequency of 1 MHz. Clearly, the demodulator output voltage is no longer a sine wave. The demodulator output voltage can be written as a Fourier series summation

$$V_{DEM}(t) = \sum_{n=0}^{\infty} (A_n \cos n\omega_m t + B_n \sin n\omega_m t). \tag{3.43}$$

Let $V_1(t)$ be the fundamental component of $V_{DEM}(t)$ equal to

$$V_1(t) = HD_1 \cos(\omega_m t + \phi) = A_1 \cos \omega_m t + B_1 \sin \omega_m t, \tag{3.44}$$

where A_1 and B_1 are given by the Fourier integrals

$$A_1 = \frac{2}{T} \int_0^T V_{DEM}(t) \cos \omega_m t \, dt, \tag{3.45}$$

and

$$B_1 = \frac{2}{T} \int_0^T V_{DEM}(t) \sin \omega_m t \, dt. \tag{3.46}$$

The amplitude of the fundamental component of the subcarrier (HD_1) is given by

$$HD_1(t) = \sqrt{A_1^2 + B_1^2}. \tag{3.47}$$

In order to investigate the effect of the frequency selective fading upon the subcarrier fundamental HD_1, and to identify good and bad propagation channels, it is necessary to transform (3.40) into the frequency domain

$$V_{DEM}(f) = K_D B(f) * \delta(f - f_m). \tag{3.48}$$

The fundamental component of the subcarrier can be written as

$$HD_1 = V_{DEM}(f = f_m). \tag{3.49}$$

In order to isolate the effects of the frequency-selective fading, it is necessary to normalize the RF input signal power. This is accomplished by ensuring that

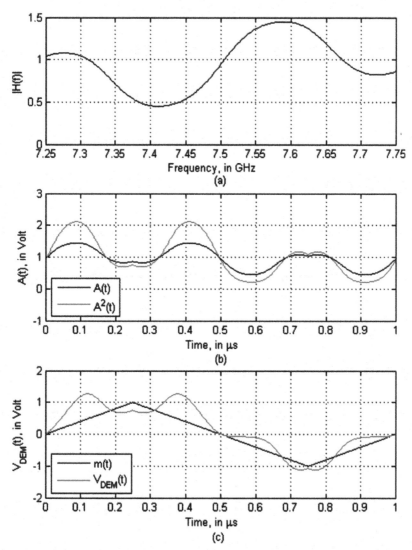

Figure 3.14 (a) Example of channel transfer function, (b) instantaneous amplitude and its square, and (c) FM demodulator output voltage.

the RMS value of $A(t)$ equals one; as a result, the average value of $B(t) = A^2(t)$, is also unity. The RMS value of the channel transfer function $H(f)$ is also one.

Given the convolution with $B(f)$, two components, $B(0)$ and $B(2f_m)$ seem to be able to influence HD_1. Due to the applied normalization, the average

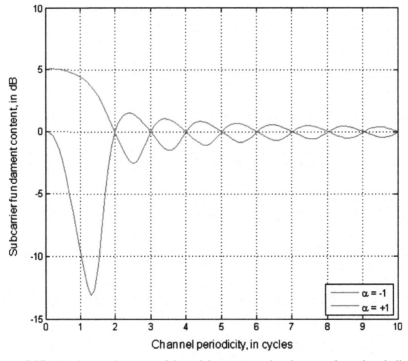

Figure 3.15 Fundamental content of demodulator output signal versus channel periodicity.

value of B(t) is constant, and so will be B(0). The only component that can affect HD_1 is the component $B(2f_m)$ at two times the subcarrier frequency. The sign of this component decides whether HD_1 increases or decreases. Channel transfer functions H(f) with a periodicity (P_C) of one cycle across the FM-UWB signal bandwidth, have a strong component $B(2f_m)$ and yield the extreme cases for the propagation. Faster channel transfer variations, i.e., multiple cycles across the FM-UWB signal bandwidth hardly affect the performance.

3.3.1.1 Best and Worst Case Propagation Channels
In order to test the hypothesis stated in the previous subsection, verification was performed using a simple single-reflection channel with variable periodicity and channel transfer function equal to

$$H(f) = 1 + \alpha \cos \left(2\pi \frac{f - f_C}{B_{RF}} P_C \right) \qquad (3.50)$$

With a channel periodicity P_C of 1, H(f) makes one cycle across the bandwidth B_{RF} (i.e., from $f_C - \triangle f$ to $f_C + \triangle f$). An α of -1 yields the best channel, and an α of $+1$ yields the worst case channel. Figure 2.26 shows the fundamental content of the demodulator output versus the channel periodicity for best and worst case channels.

It can be seen that the worst case occurs for an α of 1 and channel periodicity of 1.3, as HD_1 drops by 13 dB. Figure 3.16 shows the channel transfer function and time domain waveforms of m(t), A(t), and demodulator output voltage for this case. The reason why the worst case doesn't occur at a channel periodicity of unity is not clear and is a subject for future work.

Figure 3.17 shows the best case deterministic channel occurring for an α of -1 and $0 < P_C \ll 1$, note that $P_C = 0$, would yield a flat channel.

Comparing Figures 3.16 and 3.17 gives insight into what makes a channel good or bad for the FM-UWB radio. Values of the channel transfer function H(f) around the center frequency f_c, where m(t) = 0, hardly affect the subcarrier amplitude HD_1, whereas the value at the extremes of the FM-UWB signal bandwidth, i.e., at $f_C - \triangle f$ and $f_C + \triangle f$, where m(t) = ± 1, have a stronger impact on the subcarrier amplitude HD_1, as is clearly illustrated by Figure 3.17b.

The explanation is the following: the FM demodulator output signal is proportional to the square of the instantaneous amplitude A. When A has its maximum value while m(t) makes its zero crossing ($f_{INST} \approx f_C$), which is the case for the channel shown in Figure 3.16, the resulting subcarrier amplitude will be low. When A has its maximum values coinciding with the maxima of m(t), ($f_{INST} \approx f_c \pm \triangle f$), as is the case for the channel transfer function shown in Figure 3.17, the resulting subcarrier amplitude will be high. In the next sections, we will investigate whether this also holds for practical wireless channels as occurring in BAN applications targeted by FM-UWB radio.

3.3.2 Performance Evaluation with Statistical Channel Models

Since the receiver performance in terms of probability of error is fully determined by the subcarrier SNR, it is not necessary to implement a complete simulation chain for the FM-UWB system to investigate the effect of frequency selective fading. It is sufficient to calculate the fundamental content (HD_1)of the subcarrier signal for each channel realization and investigate its statistical properties for a large number (e.g., 1000) of channel realizations.

For each channel realization, the channel impulse response, h(t), is generated. The frequency domain channel transfer function, H(f), is obtained

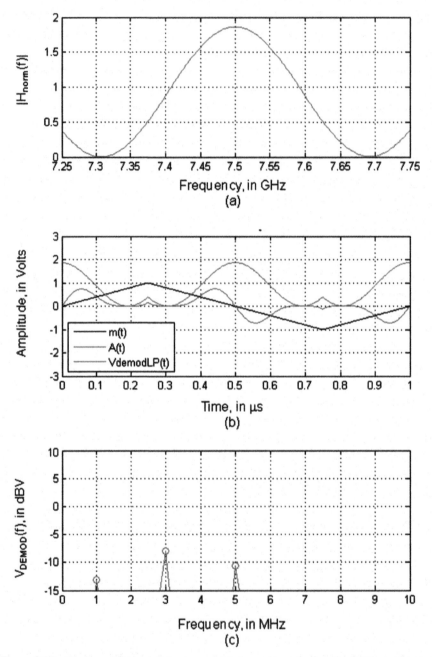

Figure 3.16   Worst case deterministic channel: (a) channel transfer function, (b) instantaneous amplitude and FM demodulator output, (c) spectrum of FM demodulator output signal.

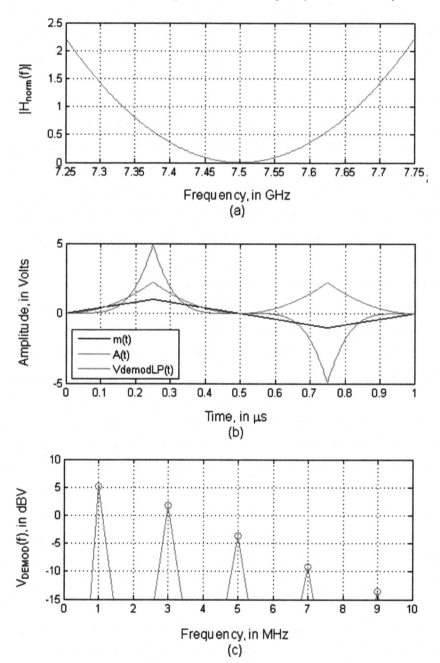

Figure 3.17 Best case deterministic channel: (a) channel transfer function, (b) instantaneous amplitude and FM demodulator output, (c) spectrum of FM demodulator output signal.

by taking the discrete Fourier transform (DFT) of the channel impulse response. Figure 3.18a shows the impulse response and Figure 3.18b shows the corresponding channel transfer function for a randomly chosen realization of a IEEE802.15.3a CM2 channel [7].

The normalized version of H(f), i.e., $H_{norm}(f)$, is used to make a fair comparison between the various channel realizations, i.e., equal RF power for all cases, distinguishing themselves only by the frequency-selective fading. Next, the time dependent amplitude of the received signal, A(t), is calculated using (3.41).

The demodulator output voltage, $V_{DEM}(t)$, can now be calculated using (3.42). Figure 3.18c shows the subcarrier signal, m(t), the time-varying amplitude, A(t), and the FM demodulator output signal, $V_{DEM}(t)$. As a last step, the fundamental of the subcarrier signal (HD_1) is calculated by taking the FFT of the demodulator output voltage as illustrated in Figure 3.18d.

The next two subsections will show the performance of the FM-UWB system with the IEEE 802.15.3a and IEEE802.15.6 channel models.

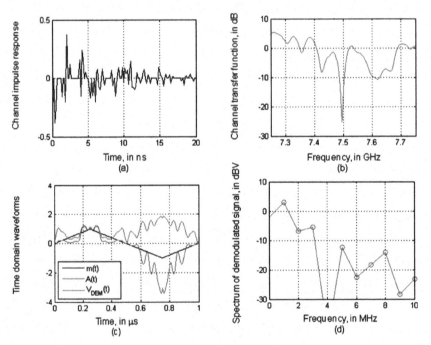

Figure 3.18   (a) Channel impulse response, (b) channel transfer function, (c) subcarrier, amplitude of receiver input signal and FM demodulator output signal, (d) spectrum of FM demodulator output signal.

3.3.2.1 Performance with 802.15.3a UWB channel models

As a first example of FM-UWB performance with frequency-selective fading, the IEEE 802.15.3a channels for indoor propagation [7] have been used. Table 3.2 presents their characteristics. These models represent channels varying from line-of-sight (LOS) to highly dispersive with delay spread equal to 25 ns.

For each of the CM1–CM4 channels, 1000 channel realizations were generated and statistics were performed using MATLAB. Table 3.3 presents the results.

The numbers show that the average influence of frequency selective multipath is equal to 0 dB, and that in 50% of the cases, multipath boosts the performance. It can also be seen that the spread between maximum and minimum values becomes lower for stronger multipath (CM4).

Variations in subcarrier level can be translated into the RF fading level, i.e., the (equivalent) receiver input power level [10], by taking the square root, or taking half the value when expressed in dB. The RF fading level defines the margin, i.e., the increase in RF signal level, that is required to obtain a given receiver BER performance with multipath. Positive values of the fading level indicate a negative margin, i.e., a virtual receiver sensitivity increase.

In the worst case CM1 channel the RF fading level is –3.6 dB, meaning that 3.6 dB more RF signal is required to obtain the same probability of error compared to the case of no frequency-selective fading. In other words, the fading margin equals 3.6 dB.

The statistics presented above confirm the good performance for FM-UWB system in the presence of frequency-selective multipath. Again, it is important to understand what constitutes good and bad channels for a

Table 3.2 IEEE802.15.3a channel characteristics

Channel	Characteristics
CM1	LOS, D < 4 m
CM2	NLOS, D < 4 m
CM3	NLOS, 4 m < D < 10 m
CM4	NLOS, α_τ = 25 ns

Table 3.3 Variations in subcarrier level, in dB

Channel	Min	Max	Average	Median
CM1	–7.16	+4.06	–0.09	+0.20
CM2	–5.83	+3.92	–0.02	+0.01
CM3	–6.07	+3.74	–0.06	+0.08
CM4	–4.91	+3.09	–0.04	+0.03

FM-UWB communication system. Inspection of the normalized channel transfer function of the best and worst cases of the CM1 and CM4 channels as shown in Figures 3.19 and 3.20 is helpful.

Although the channels investigated have different properties (e.g., the coherence bandwidth of the CM4 channel is much lower than of the CM1 channel), there appears to be a common characteristic for good and bad channels, illustrated by the dashed line in Figures 3.19 and 3.20. Values of the channel transfer function $H(f)$ around the center frequency f_c, where $m(t)$ is zero, hardly affect the subcarrier amplitude HD_1, whereas the value at the

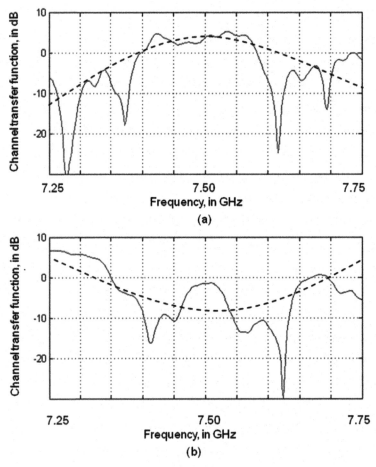

Figure 3.19   Channel transfer function for (a) worst and (b) best CM1 channel.

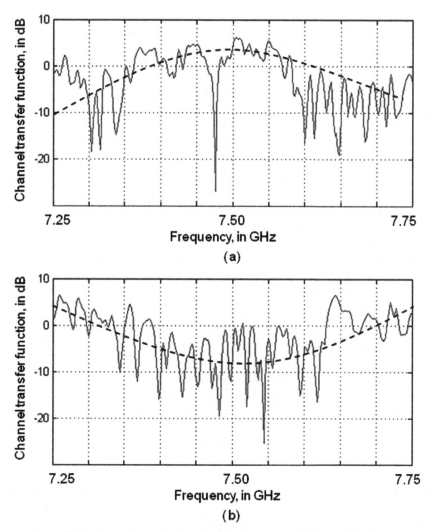

Figure 3.20 Channel transfer function for (a) worst and (b) best CM4 channel.

extremes of the FM-UWB signal bandwidth, i.e., at $f_C - \triangle f$ and $f_C + \triangle f$, where m(t) is ± 1, have a stronger impact on the subcarrier amplitude HD_1.

These results are consistent with the results already found for the deterministic single-reflection channel and confirm that a notch in the channel transfer function in the vicinity of the center frequency has little impact on performance. Such a notch may also arise from a bandstop filter used to mitigate in-band interference. It is shown in Appendix B that, when the notch is not too wide (less than 20% of the FM-UWB bandwidth), the impact on performance is acceptable, even when the notch doesn't occur at the center frequency.

3.3.2.2 Performance with IEEE802.15.6 UWB BAN channel models

The IEEE802.15 Task Group 6 (IEEE802.15.6) is currently developing a BAN communication standard [11], and specific BAN channel models have been developed [9]. Body surface to body surface communication is modeled by the CM3 channel. Transmission from the body surface to external devices is modeled by the CM4 channel.

Figure 3.21 shows MATLAB simulation results of the PDF of the RF fading level, i.e., the equivalent receiver input power for 4000 realizations of the IEEE CM3 and CM4 channels. A fading level of 0 dB corresponds to the

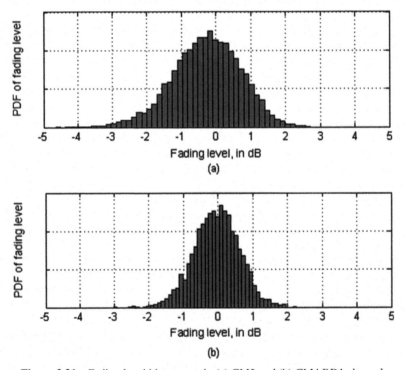

Figure 3.21 Fading level histograms in (a) CM3 and (b) CM4 BD1 channel.

case of no fading relative to the mean signal level. The mean and median values of the fading distribution were both found to be 0 dB, meaning that 50% of the time performance improves and 50% of the time performance degrades. This is clearly illustrated by the histograms of Figure 3.21.

Figure 3.22 shows the cumulative density function (CDF) of the fading level for the FM-UWB signal. It can be seen that 99% of the time the fading level is above –2.8 dB for the CM3 channel, and above –1.7 dB for the CM4 channel. This means that 2.8 dB of fading margin is required to achieve 99% availability in the CM3 channel, and only 1.7 dB of fading margin is required in CM4. This compares favorably to a narrowband radio, which requires 20 dB higher received power for 99% availability [10]. This is illustrated in Figure 3.23, which shows the fading level in a CM3 channel for both the FM-UWB and a narrowband FSK signal. The arrow shows the 17 dB of difference in fading level for 99% availability arising from the diversity provided by the ultra-wideband transmit signal over the frequency selective multipath fading channel as defined in CM3 and CM4. Importantly, this is achieved without additional receiver complexity for FM-UWB given the "narrowband" signal detection in the subcarrier.

3.4 FM-UWB Performance with Interference

Interference mitigation is one of the major challenges in today's UWB receivers. In-band interference is present from other UWB radios and WiMAX communication systems, out-of-band interference originates from WLANs and cellular systems. Two important parameters are the interferer's modulation scheme and modulation bandwidth.

In the wideband FM demodulator, the interfering signal is multiplied with a delayed version of itself and with the wanted FM-UWB signal. The first case yields combined AM-FM demodulation of the interfering signal. When the interferer is at the center frequency of the FM demodulator, AM rejection is high and only the FM of the interferer will be present at the demodulator output. When a frequency offset is present, the AM rejection will be much lower and additional actions may be required to mitigate the interference. WiMAX, WiMedia and impulse radio signals are examples of interferers with envelope modulation (AM) characteristics. An example of an FM interferer could be the third harmonic of an FSK signal in the 2.4 GHz ISM band (e.g., a Bluetooth system).

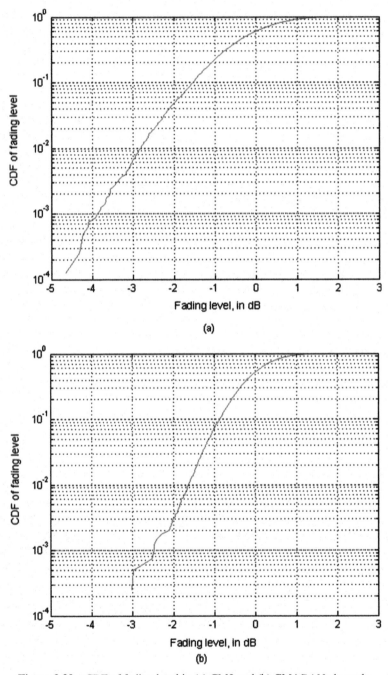

Figure 3.22   CDF of fading level in (a) CM3 and (b) CM4 BAN channels.

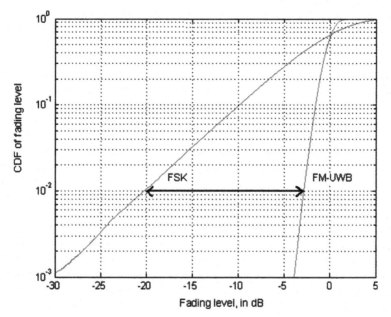

Figure 3.23 CDF of fading level in CM3 channel for narrowband FSK and 500 MHz wide FM-UWB signals.

The second case, i.e., multiplication of the interfering signal with the FM-UWB signal, always yields a wideband signal and spreads the interferer signal over a large bandwidth.

As a simple case, we assume a CW jammer at the center frequency of the FM-UWB signal. This case is like the multiple-access interference for the 2-user case as presented in Section 3.2.1, but with an unmodulated interferer. The residue signal V_{NO} is the result of the CW signal and the FM-UWB signal down-converting each other yielding a FM signal centered at zero. Its single-sided power spectral density is flat and has a bandwidth equal to half the RF bandwidth of the FM-UWB signal ($B_{RF} = 2\triangle f$). Only the part of the residue falling within the subcarrier bandwidth contributes to the degradation of the subcarrier SNR. The receiver processing gain, G_P, is defined in (3.11) as the ratio of the RF and subcarrier bandwidth.

The subcarrier SNR is given by

$$SNR_{SUB}\frac{B_{RF}}{2B_{SUB}}SNR_{DEM} = G_P\frac{SIR}{8}. \tag{3.51}$$

The effect of a CW interferer is 3 dB stronger than of a FM-UWB interferer.

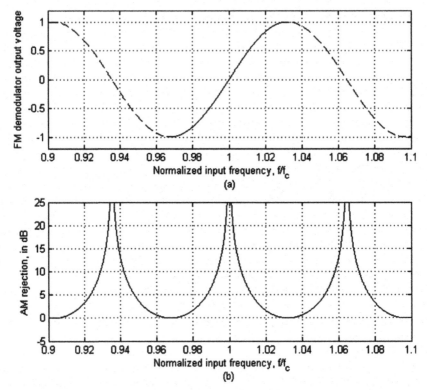

Figure 3.24   Example of (a) FM demodulator output voltage and (b) AM rejection versus normalized frequency.

The most harmful result of interference in an FM-UWB system is direct demodulation of narrowband AM interferers.

3.4.1 Out-of-Band Interference

Out-of-band interference can be reduced significantly by passive filtering prior to the UWB receiver. Measurements performed on a commercially available passive filter for the UWB low band show an insertion loss of 1 dB at 4.5 GHz and 23 dB attenuation at 5.25 GHz. This compact filter can be easily inserted between the antenna and UWB receiver. Antenna tuning may also help to mitigate interference, since they can be designed to have notches in their frequency response. In [12] a UWB antenna with a 10 dB notch in its frequency response at 5.2 GHz is presented. Examples of UWB antenna where the notch frequency can be electronically tuned, are described in [13] and [14].

3.4.2 In-Band Interference

In-band interference can be mitigated by receiver processing gain, RF filtering prior to the FM demodulator, and (as a final resort) by applying detect-and-avoid (DAA) techniques [15].

The receiver processing gain is most efficient for wideband interferers like WiMedia. Wideband interfering signals also yield a wideband signal after the FM demodulator. However, only the part of the demodulated signal falling inside the subcarrier bandwidth is harmful. The worst case interference is constituted by a signal modulated by a single sine wave at the subcarrier frequency of the wanted signal. Such an interfering signal is not very likely to occur. It could arise from an FM-UWB user in a neighboring piconet that uses the same subcarrier frequency. In-band impulse radio UWB systems may operate at a pulse repetition rate of 1 MHz [16]. However, due to the use of spreading codes at 16 times the bit rate, the energy is not concentrated around that frequency, but spread over a larger bandwidth.

Medium bandwidth interference, typically WiMAX communication systems with a −20 dB bandwidth up to 21 MHz [17], will yield high power inside the subcarrier bandwidth signal after the FM demodulator.

In-band interference whose bandwidth is smaller than the FM-UWB signal bandwidth can be attenuated by a notch filter. Although the signal power of the FM-UWB signal slightly decreases, the signal-to-interference ratio (SIR) is increased. Active bandstop filtering can also be implemented in the receiver low-noise amplifier [18]. For the wanted FM-UWB signal, filtering prior to the FM demodulator is indistinguishable from a frequency notch caused by frequency-selective fading. In Appendix B, it is shown that the effect of such a notch upon the subcarrier signal amplitude can be negligible.

3.4.3 AM Demodulation and AM Rejection by the FM Demodulator

Given the fact that most in-band interference is AM, it is worthwhile to determine the AM sensitivity of the wideband FM demodulator. Consider the following AM input signal $V_{AM}(t)$ of frequency f_c and time dependent amplitude $A(t)$, given by

$$V_{AM}(t) = A(t)\sin(\omega_c t). \qquad (3.52)$$

The multiplier output signal (V_{DEM}) can be written as

$$V_{DEMOD}(t) = A(t)A(t-\tau)\sin(\omega_c t)\sin(\omega_c(t-\tau)). \qquad (3.53)$$

The lowpass filtered version of this signal can be written as

$$V_{DEM}(t) = \frac{1}{2}A(t)A(t-\tau)\cos(\omega_c\tau)$$

$$\approx \frac{1}{2}A^2(t)\cos(\omega_c\tau). \tag{3.54}$$

The approximation is valid for amplitude variations whose bandwidth is much smaller than $1/\tau$. An amplitude modulated signal with sinusoidal modulation at modulation frequency f_m has a time-varying envelope $A(t)$ given by

$$A(t) = A_1(1 + m\cos(\omega_m t)). \tag{3.55}$$

The lowpass filtered demodulated signal V_{DEM} is equal to

$$V_{DEM}(t) = A_1^2 \left[1 + \frac{m^2}{2} + 2m\cos(\omega_m t) + \frac{m^2}{2}\cos(2\omega_m t) \right]\cos(\omega_c\tau), \tag{3.56}$$

which for low modulation depth m, can be approximated by

$$V_{DEM}(t) \approx A_1^2 \left[1 + 2m\cos(\omega_m t) \right]\cos(\omega_c\tau). \tag{3.57}$$

This implies that the AM sensitivity is zero at the operating points chosen for FM demodulation as given by (2.9) and has its maximum values in between, where the term $\cos(\omega_c\tau)$ has its extreme values. As a result, the delay line demodulator provides strong AM rejection for narrowband signals centered on those operating points, as illustrated in Figure 2.35 for a delay line demodulator with useful bandwidth of 500 MHz and a demodulator center frequency fc = 7.5 GHz, requiring a delay time $\tau = 31/(4f_C)$.

3.5 Conclusions

This chapter has presented the theoretical performance of the FM-UWB communication system under various conditions.

Performance with AWGN shows a link span, that it is well suited for short range BAN applications. Up to 17 meters is obtained under free-space propagation conditions for a 500 MHz wide FM-UWB signal at 250 kbps and assuming a 5 dB noise figure for the receiver LNA.

The performance with frequency-selective multipath, typically the IEEE UWB and BAN channel models has been investigated. The mean and median values of the fading distribution were both found to be 0 dB, meaning that 50% of the time performance improves and 50% of the time performance degrades.

It was found that 99% of the time the fading level is above –2.8 dB for the CM3 BAN channel, and above –1.7 dB for the CM4 BAN channel. This means that 2.8 dB of fading margin is required to achieve 99% availability in the CM3 channel, and only 1.7 dB of fading margin is required in CM4. This compares favorably to a narrowband radio, which requires 20 dB higher received power for 99% availability.

Multiple-access interference is generated in the wideband FM demodulator and attenuated by the receiver processing gain which is equal to the ratio of the RF and subcarrier bandwidth. E.g., in a 100 kbps system with a RF bandwidth of 500 MHz, a processing gain of 34 dB is available. This allows for a FM-UWB interferer that is 24 dB stronger than the wanted signal, a value that is realistic for a short-range LDR PAN scenario. A CW jammer 21 dB stronger than the wanted signal can be tolerated.

The maximum number of users in a FM-UWB communication system based upon subcarrier FDMA is dependent on the bit rate. In theory, a 10 kbps system with a RF bandwidth of 1 GHz can accommodate 50 equal power users, whereas a 100 kbps system accommodates 15 users, provided that the subcarrier frequency is chosen sufficiently high. It was shown in Chapter 2 that the use of subcarrier frequencies above 2 MHz requires reduction of transmit power, which lowers the link span. A trade-off needs to be made between the number of simultaneous users and the link span. The result may be that the subcarrier frequency range is restrained.

The capacity of a low-complexity FM-UWB system using binary FSK as subcarrier modulation scheme, is equal to half the available subcarrier frequency range. When the optimum subcarrier frequency range, i.e., 1–2 MHz is used, this yields a capacity of 500 kbps.

The effect of in-band interference is mitigated by the receiver processing gain. In-band interference whose bandwidth is smaller than the FM-UWB signal bandwidth can be attenuated by a notch filter. The effect of such a notch upon the receiver sensitivity is negligible. Out-of-band interference can be reduced significantly by passive filtering prior to the UWB receiver. Antennas can also be designed to have notches in their frequency response.

References

[1] John G. Proakis, *Digital Communications*, Third Edition, New York: McGraw-Hill, 1995.
[2] M. H. L. Kouwenhoven, *An analysis of the quadrature and mathematical demodulator in the presence of noise*, Internal Report, Electronics

Research Laboratory, Delft University of Technology, Delft, the Nether-lands, November 1995.

[3] M. H. L. Kouwenhoven, *High-Performance Frequency-Demodulation Systems*, PhD thesis, Technische Universiteit Delft, 1998.

[4] John F. M. Gerrits, John R. Farserotu, and John R. Long, "Multipath Behavior of FM-UWB Signals," in *proc. ICUWB2007*, pp. 162–167.

[5] Kaveh Pahlavan, and Allen. H. Levesque, *Wireless Information Networks*, New York: John Wiley and Sons, 1995.

[6] R. Vaughan and J. B. Andersen, *Channels, Propagation and Antennas for Mobile Communications*, London: The Institution of Electrical Engineers, 2003.

[7] J. R. Foerster, M. Pendergrass and A. F. Molisch, "A Channel Model for Ultrawideband Indoor Communications," *Proc. International Symposium on Wireless Personal Multimedia Communication*, October 2003.

[8] Andreas F. Molisch, *IEEE 802.15.4a channel model subgroup final report*, Doc.: 15-04-0535-00-004a-channel-model-final-report. [Online]. Available: https://mentor.ieee.org/802.15/dcn/04/15-04-0535-00-004a-tg4a-channel-model-final-report.pdf

[9] Kamya Yekeh Yazdandoost, *Channel Model for Body Area Network (BAN)*, IEEE 802.15-08-0780-10-0006, 14 July 2010. [Online]. Available: https://mentor.ieee.org/802.15/dcn/08/15-08-0780-10-0006-tg6-channel-model.pdf

[10] M. Wolf, G. Del Galdo, M. Haardt, "Performance evaluation of ultrawide band compared to wireless infrared communications," *Proceedings ECWT2005*, pp. 193–196.

[11] Body Area Networks (BAN), IEEE 802.15 WPAN[TM] Task Group 6, [Online]. Available: http://www.ieee802.org/15/pub/TG6.html

[12] J. Kim, C. S. Cho, and J. W. Lee, "5.2-GHz notched ultra-wideband antenna using slot-type SRR," *Electronics Letters*, vol. 42, no. 6, pp. 315–316, March 2006.

[13] E. Antoniono-Daviu, M. Cabedo-Fabrés, M. Ferrando-Bataller, and A. Vila-Jimenez, "Active UWB antenna with tunable band-notched behaviour," *Electronics Letters*, vol. 43, no. 18, pp. 959–960, August 2007.

[14] Julien Perruisseau-Carrier, Pablo Pardo-Carrera, and Pavel Miskovsky, "Modeling, Desing and Characterization of a Very Wideband Slot Antenna with Reconfigurable Band Rejection," *IEEE transactions on antennas and propagation*, vol. 58, no. 7, pp. 2218–2226, July 2010.

[15] J. Lansford and D. Shoemaker, "Technology tradeoffs for a worldwide UWB transceiver," in *Proc. ICUWB2007*, pp. 259–263.

[16] IEEEP802.15.4a/D4 (Amendment of IEEE Std 802.15.4), Part 15.4: *Wireless medium access control (MAC) and physical layer (PHY) specifications for low-rate wireless personal area networks (LRWPANs)*, July 2006.

[17] IEEE Std 802.16-2004, *IEEE Standard for Local and metropolitan area networks, Part 16: Air Interface for Fixed Broadband Wireless Access Systems*, [Online]. Available: http://standards.ieee.org/getieee802/down load/802.16-2004.pdf

[18] G. Yao, Y. J. Zheng, and B. L. Ooi, "0.18 mm CMOS dual-band UWB LNA with interference rejection," *Electronics Letters*, vol. 43, no. 20, pp. 1096–1098, September 2007.

4

FM-UWB Transmitter Implementation

This chapter shows how the FM-UWB transmitter can be implemented. Section 4.1 addresses the subcarrier generation using direct digital synthesis (DDS) techniques [1]. Section 4.2 presents the RF signal generation using a mix of analog and digital techniques. The FM-UWB signal is generated by a free-running RF VCO that can be calibrated by a PLL frequency synthesizer. The influence of non-idealities, like VCO tuning curve non-linearity and oscillator phase noise, on the FM-UWB radio performance is analyzed. Implementation examples of the output amplifier are presented. Section 4.3 presents conclusions.

4.1 DDS-based Subcarrier Signal Generation

The subcarrier generation is implemented using a DDS. The DDS comprises a phase accumulator and a phase-to-amplitude converter as shown in Figure 4.1.

The N-bit wide output word of the phase accumulator (P) represents the phase ϕ of the generated waveform. P has values $0 < \phi < 2^N - 1$ corresponding to phase values $0 < \phi < (2\pi - \Delta\phi)$, where $\Delta\phi$ is the phase resolution.

$$\phi = P\Delta\phi = \frac{P}{2^N}2\pi$$
$$\Delta\phi = \frac{2\pi}{2^N}$$

(4.1)

In the absence of FSK modulation, a signal of fixed frequency (f_{SUB}) is generated. At each clock pulse, the phase register is incremented by ΔP which corresponds to a phase increment equal to $N_{TX}\Delta\phi$ where

$$N_{TX} = 2^N\frac{f_{SUB}}{f_{CLK}}.$$

(4.2)

The phase evolves like a sawtooth at the subcarrier frequency. The phase register overflows for values above $2^N - 1$ and thus counts modulo 2^N. The

73

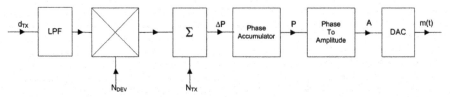

Figure 4.1   Block diagram of the transmitter DDS.

frequency resolution (δf) provided by a DDS with N-bit phase accumulator operating at a clock frequency f_{clk} is given by

$$\delta f = \frac{f_{CLK}}{2^N}. \tag{4.3}$$

The output word of the phase accumulator (P) is next converted into a word (A) that represents the instantaneous amplitude of the subcarrier signal and drives the DAC. This word is truncated to M bits, where M < N. The value of M is determined by the maximum allowed spurious level [1]. A value M = 10 yields spurious levels 60 dB below the main signal. When a DDS generates a sine wave, the phase-to-amplitude conversion is usually implemented by a look-up table. When a triangular wave is generated, simple combinatorial logic can be used, as shown in Figure 4.2, resulting in reduced hardware complexity [2]. The most significant bit of the phase accumulator word (P_{N-1}) is used to invert the next M bits $(P_{N-2} \ldots P_{N-M-1})$.

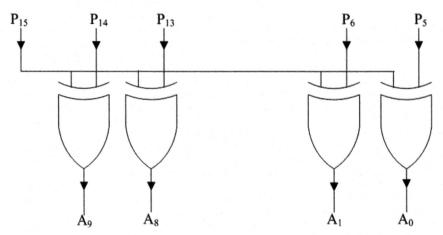

Figure 4.2   Phase-to-amplitude conversion circuit for N = 15 and M = 10 [3].

The DAC output signal is lowpass filtered to attenuate the aliasing components and to obtain a smooth waveform. This filter has a 2nd order Butterworth characteristic. It can be implemented as an analog filter or as a linear interpolation DAC [3].

4.1.1 DDS Clock Frequency

The clock frequency is chosen in such a way that for the highest generated subcarrier frequency value, the signal waveform closely resembles a triangle with uniform probability density function. This requires the presence of the fundamental and third, fifth and seventh harmonics. It is also useful to choose the clock frequency as a power of 2 times the data rate in order to simplify the clock divider chain that generates the clock signals for the digital baseband. For a maximum subcarrier frequency of 2 MHz, a clock frequency of 32 MHz meets the requirements. With N = 16 and f_{CLK} = 32 MHz, a frequency resolution of 488 Hz is obtained.

4.1.2 FSK Modulation of the DDS

FSK modulation is implemented by making the phase increment (ΔP) dependent on the transmit data (d_{TX}) and the subcarrier deviation (Δf_{SUB}). When the transmit data is represented by a sequence of binary symbols (e.g., +1 and −1), the phase increment is given by

$$\Delta P = N_{TX} + N_{DEV} H_{LPF}(d_{TX}), \tag{4.4}$$

where $H_{LPF}(d_{TX})$ represents the lowpass filtered data signal, and N_{DEV} is given by

$$N_{DEV} = 2^N \frac{\Delta f_{SUB}}{f_{CLK}}. \tag{4.5}$$

4.1.3 Data Lowpass Filtering

The digital data is upsampled by a factor 8 and lowpass filtered by a Gaussian filter (denoted by LPF in Figure 4.1) with bandwidth-symbol duration product (BT) equal to 0.8 prior to FSK modulation to reduce the sidelobe interference in adjacent subcarrier channels. In order to reduce the complexity, easy-to-implement rounded values (4-bit) are used for the filter step response, as shown in Table 4.1. Figure 4.3 shows the effect of the filter in the time domain.

Table 4.1 Step response of data lowpass filter

Data Transition	Step Response Values
0 > 0	[−1, −1, −1, −1, −1, −1, −1, −1]
0 > 1	[−1, −1, −¾, −¼, ¼, ¾, 1, 1]
1 > 0	[1, 1, ¾, ¼, ¼, ¾, −1, −1]
1 > 1	[1, 1, 1, 1, 1, 1, 1, 1]

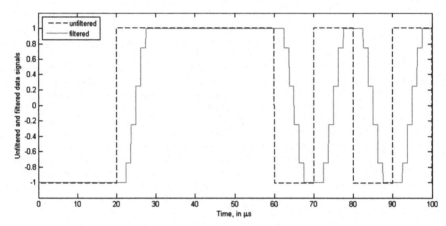

Figure 4.3 Time domain response of the data lowpass filter.

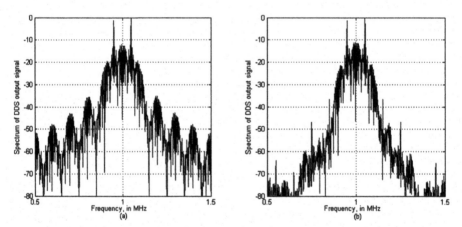

Figure 4.4 Frequency spectrum of DDS output signal (a) without filtering and (b) with filtering.

The rounding has negligible effect on the spectral purity of the FSK signal. Figure 4.4a shows the spectrum of the FSK signal (f_{SUB} = 1 MHz, Δf_{SUB} = 50 kHz) without data prefiltering and Figure 4.4b shows the spectrum with data prefiltering.

4.1.4 DDS Complexity and Power Consumption

The DDS from Figure 4.1 (not including the DAC) with a 16-bit phase accumulator and 10-bit amplitude as described in [3] requires 3167 logical gates and has a chip area of 0.12 mm^2, when realized with standard digital cells in 0.13 μm CMOS technology. The simulated power consumption of the DDS at 32 MHz clock frequency is 400 μW and the simulated power consumption of the interpolating DAC is 350 μW [4].

4.2 RF Signal Generation

The FM-UWB signal is generated by a free-running RF VCO that can be calibrated by a PLL frequency synthesizer as shown in Figure 4.5. This frequency synthesizer ensures the correct center frequency of the UWB signal and the corresponding VCO voltage is measured with an ADC and memorized in an EEPROM. Modulation of the RF VCO with the subcarrier signal occurs in open loop mode. The VCO voltage value that was stored during calibration is now generated by a DAC and the subcarrier signal is added to it.

The VCO output is the FM-UWB signal which is fed to the output amplifier (OA) driving the antenna. The VCO output signal is also fed into a fixed ratio prescaler ($N_{FIX} = 256$) that reduces the VCO output frequency (6–9 GHz) to a lower frequency (23–35 MHz) that is compatible with the programmable divider hardware (N_{VAR}). The RF center frequency is given by

$$f_{RF} = N_{VAR}\, N_{FIX}\, f_{REF}. \tag{4.6}$$

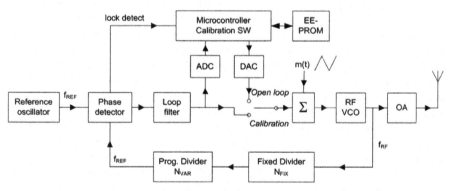

Figure 4.5 Block diagram of RF signal generation.

With a reference frequency equal to 250 kHz, the center frequency of the UWB signal will have a resolution of 64 MHz.

As the frequency synthesizer doesn't need to operate continuously, its contribution to the average power consumption of the transmitter can be made negligibly small. An initial calibration is performed the first time the transmitter is powered up and after that at low duty cycle, e.g., 1%. The required calibration strategy for reliable operation is subject of further investigations. Calibration could be performed when temperature or supply voltage changes.

4.2.1 Oscillator Type

An oscillator is an active electrical circuit that can generate periodic waveforms out of constants [5]. This short definition is illustrated in Figure 4.6. DC energy from the power supply is transformed into the time varying oscillator output signal, A(t).

In an ideal oscillator circuit, the oscillation frequency depends only on the constants and not on the active part. In a practical oscillator circuit, the active part does have an influence the oscillation frequency. Part of the frequency determining constants may be constituted by the active circuit's parasitic capacitances.

The oscillator is characterized by its tuning range, power consumption frequency stability. The short term frequency fluctuations, referred to as phase noise, determine together with the quality factor (Q) of the resonator the required resonator power and DC power consumption.

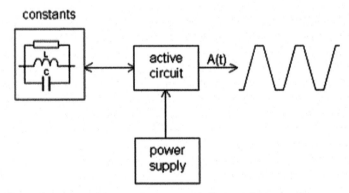

Figure 4.6   An oscillator generates a periodic waveform out of constants.

As will be shown in Section 4.2.3, the phase noise specification of the transmit VCO in the FM-UWB transmitter is not severe. However, since it is specified at 1 to 2 MHz offset from the carrier, 1/f noise could be a problem. In order to end up with low power consumption, it is advantageous to use a harmonic oscillator. Low noise oscillators use resonators with high quality factor to determine the oscillation frequency [6].

4.2.2 VCO Tuning Curve Non-Linearity

The effect of non-linearities of the RF VCO tuning curve can be modeled as shown in Figure 4.7. An ideal VCO is driven by a predistorted version m′(t) of the modulating signal m(t) where

$$m'(t) = \Phi(m(t)). \tag{4.7}$$

The power spectral density of the wideband FM signal V(t) has the shape of the probability density function of the predistorted modulating signal m′(t) [7]. It was found experimentally that the tuning curve of certain VCOs (using junction varactor diodes) can be approximated by a hyperbolic tangent non-linearity. Figure 4.8a shows the tuning curve of the VCO used in the hardware prototype of the FM-UWB radio [8]. The VCO tuning voltage V_T is between 0 and 1.2 V. Figure 4.8b shows VCO gain versus tuning voltage, and Figure 4.8c shows tuning gain versus output frequency. The three dots correspond to frequencies of 6.25, 7 and 7.25 GHz.

Figure 4.9 shows MATLAB simulation results of what happens when this VCO is used to generate 500 MHz wide FM-UWB signals at center frequencies of 6.25, 7.00 and 7.75 GHz. Figure 4.9a shows the 1 MHz triangular subcarrier signal m(t), Figure 4.9b shows the instantaneous frequency (f_{INST}), of the FM-UWB signal, which is similar in shape to m′(t). Figure 4.9c shows the histogram of the instantaneous frequency which has the same shape as the power spectral density of the FM-UWB signal.

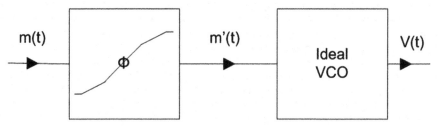

Figure 4.7   Model of VCO with tuning curve non-linearity.

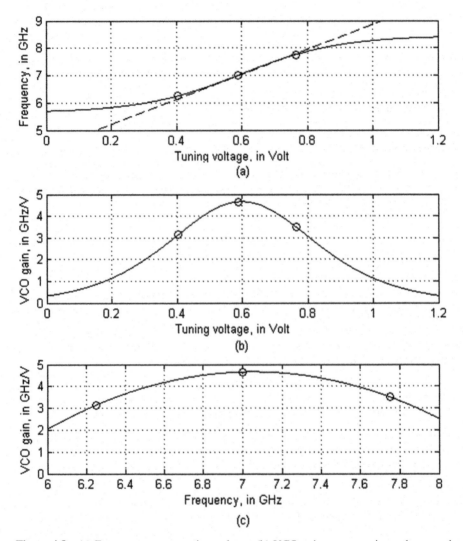

Figure 4.8 (a) Frequency versus tuning voltage, (b) VCO gain versus tuning voltage, and (c) VCO gain versus VCO frequency [8].

Only a certain portion of the VCO tuning curve is used for the generation of each FM-UWB signal. It can be shown that the normalized power spectral density (normalized to its value at f_C) is given by

$$PSD(f) = \frac{K_{VCO}(f_C)}{K_{VCO}(f)} \qquad (4.8)$$

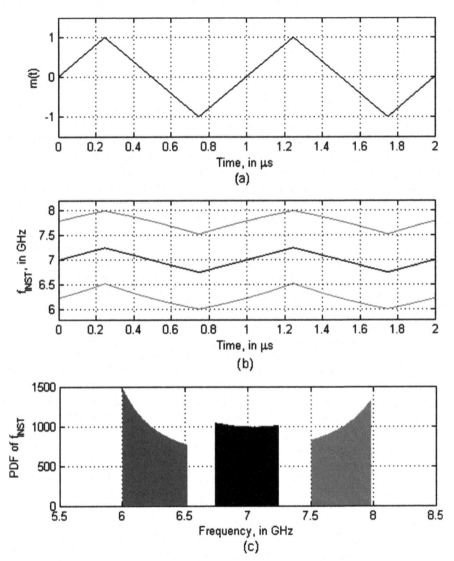

Figure 4.9 (a) Subcarrier signal, (b) instantaneous frequency of FM-UWB signal, (c) PSD of FM-UWB signal.

It is the lowest value of the VCO gain that determines the peaking of the PSD. When the peaking in dB should stay below a value N_{PEAK}, it is required that

$$K_{VCO,MIN} > 10^{-\frac{N_{PEAK}}{10}} K_{VCO}\left(f_C\right) \qquad (4.9)$$

When we allow for 1 dB of peaking of the PSD of the FM-UWB signal, the VCO gain at the band edges should not be lower than 80% of the value at the VCO center frequency.

4.2.3 RF Oscillator Phase Noise

The FM-UWB transmitter signal is not only modulated by the subcarrier signal, but also carries random frequency variations due to the RF oscillator phase noise. The spectrum $S_P(f)$ of the random phase variations is related to the single-sided power spectral density L(f) of the oscillator signal. The latter is what one observes on a spectrum analyzer. As long as the small-angle criterion is satisfied, it holds that for an offset f from the carrier

$$S_P(f) = 2L(f)[\text{rad}^2/\text{Hz}] \qquad (4.10)$$

The spectrum $S_F(f)$ of the random frequency variations is given by

$$S_F(f) = f^2 S_P(f) = 2f^2 L(f)[\text{Hz}^2/\text{Hz}] \qquad (4.11)$$

since frequency is the time derivative of the phase. The RMS value of the frequency variations Δf_{NOISE} in the bandwidth $f_1 - f_2$ is given by

$$\Delta f_{NOISE} = \sqrt{\int_{f_2}^{f_2} S_F(f) df} = \sqrt{\int_{f_2}^{f_2} 2f^2 L(f) df} [Hz] \qquad (4.12)$$

The random frequency variations are demodulated by the wideband FM demodulator in the FM-UWB receiver. Phase noise with $1/f^2$ slope yields white noise at the demodulator output. The RMS value of the frequency variations Δf_{NOISE} within the subcarrier bandwidth $[f_{SUB} - \frac{1}{2}B_{SUB}, f_{SUB} - \frac{1}{2}B_{SUB}]$ can be shown to be approximately equal to

$$\Delta f_{NOISE} \approx f_{SUB} \sqrt{2L(f_{SUB}) B_{SUB}} \qquad (4.13)$$

For a FM-UWB signal with RF deviation Δf, (RMS value $\frac{1}{2}\sqrt{2}\Delta f$) due to the subcarrier, the finite subcarrier SNR due to the signal's own phase noise SNR_{SUBPN} then equals

$$SNR_{SUBPN} = \frac{\Delta f^2}{4L(f_{SUB}) f_{SUB}^2 B_{SUB}} \qquad (4.14)$$

For a 250 kbit/s system with Δf = 250 MHz, f_{SUB} = 1 MHz, B_{SUB} = 500 kHz and a RF oscillator with L (1 MHz) = 1×10^{-8} Hz^{-1} (–80 dBc/Hz), this yields SNR_{SUBPN} = 65 dB.

This phase noise level (100 kHz RMS deviation in the subcarrier bandwidth) will not jeopardize the demodulation of the FSK modulated subcarrier.

However, in a multi-user system a stronger (unwanted) FM-UWB signal will make demodulation impossible. E.g., a 30 dB stronger interferer signal (SIR = –30 dB), e.g., with f_{SUB} = 1 MHz, Δf = 250 MHz, will result in 50 dB higher noise level and lower SNR_{SUB} to 5 dB, yielding significantly higher probability of error.

The SNR due to phase noise of a stronger interferer is given by

$$SNR_{SUBPNINT} = SIR^2 \frac{\Delta f^2}{4L\left(f_{SUB}\right) f_{SUB}^2 B_{SUB}}. \qquad (4.15)$$

Note that Δf is the deviation of the FM-UWB signal and not the offset from the carrier. Given the SIR value, it is also possible to calculate the maximum $L(f_{SUB})$ allowed for the TX.

Solving for $L(f_{SUB})$ in (4.15) yields

$$L\left(f_{SUB}\right) = SIR^2 \frac{\Delta f^2}{4SNR_{SUBPNINT} f_{SUB}^2 B_{SUB}}. \qquad (4.16)$$

A reasonable value for the transmitter phase noise is when for a given SIR value, the subcarrier SNR degradation caused by the phase noise of the stronger interferer is much lower (10 dB) than the degradation caused by the by multiple-access interference (see Section 3.2.1). With SIR = 0.01 (–20 dB), a subcarrier SNR of 23 dB is obtained for a phase noise value of –80 dBc/Hz at 1 MHz offset.

4.2.4 Oscillator Resonator Power and DC Power Consumption

Efforts have been made to improve phase noise modeling and to link phase noise to active device parameters [9–11].

In [11], it is shown that in a double switch pair MOS oscillator with parallel resonant circuit L-C-R_p, as shown in Figure 4.10, with tank voltage amplitude A, resonator power P_{res}, tank voltage amplitude A and supply current I_B are given by

$$P_{res} = \frac{A^2}{2R_p}, \qquad (4.17)$$

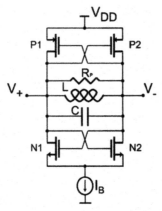

Figure 4.10   Double switch pair MOS oscillator [11].

$$A = \frac{4}{\pi} I_B R_p, \tag{4.18}$$

$$I_B = \frac{\pi}{4} \frac{A}{R_p}. \tag{4.19}$$

Phase noise at an offset $\omega (\omega = 2\pi f)$ from the carrier is given by

$$
\begin{aligned}
L\left(\Delta\omega\right) &= 10\log_{10}\left[\left(1 + \frac{\gamma_N + \gamma_P}{2}\right)kT\frac{2R_p}{A^2}\left(\frac{kT}{R_pC^2A^2\omega^2}\right)\right] \\
&= 10\log_{10}\left[2\left(1 + \frac{\gamma_n + \gamma_p}{2}\right)kT\frac{2R_p}{A^2}\left(\frac{1}{4C^2R_p^2\omega^2}\right)\right] \quad (4.20) \\
&= 10\log_{10}\left[(2 + \gamma_n + \gamma_p)\frac{kT}{P_{res}}\left(\frac{\omega_0}{2Q\omega}\right)^2\right]
\end{aligned}
$$

Where γ_n and γ_p are MOS transistor parameters with their ideal long-channel value equal to 2/3. In a practical oscillator circuit, parasitic capacitances are present at the tank nodes and performance is degraded.

The required resonator power, for a practical oscillator realization, assuming $1/f^2$ roll-off for the phase noise at an offset equal to the subcarrier frequency (1–2 MHz), is obtained from (4.20) as

$$P_{res} = (2 + \gamma_n + \gamma_p)\,kT\left(\frac{f_0}{2Qf}\right)^2 10^{-\frac{L(f_{SUB})}{10}} \tag{4.21}$$

For a 7.5 GHz oscillator with phase noise of –80 dBc/Hz at 1 MHz offset, and $\gamma_n = \gamma_p = 0.67$ [11], a resonator quality factor of 10, it follows a minimum resonator power of 200 nW.

What does this imply for the power consumption? Let's take a look at a practical oscillator with L = 2.12 nH (j100 Ω), C = 212 fF (–j100 Ω), R_p = 1 kΩ, this yields A = 20 mV, I_B = 16 uA. This looks nice, however, an amplitude of only 22 mV won't let the MOS transistors switch. It takes an amplitude 10 times higher to do so: A = 220 mV, I_B = 160 uA. Higher impedance of the parallel resonant tank circuit could solve the problem. However, this implies the use of a higher inductance and lower capacitance value. The minimum capacitance is the parasitic capacitance of the transistors and the inductance. Therefore, the calculated theoretical minimum resonator power can't always be reached in a practical circuit realization.

4.2.5 Output Amplifier Circuit Realizations

This section presents examples of output amplifiers for FM-UWB. A FM-UWB signal of –15 dBm corresponds to a voltage of 112 mV peak-to-peak in a 50 Ω load. This low value enables the use of a low supply voltage for both the oscillator and output amplifier.

A classical approach for implementing the output amplifier is the use of cascaded CMOS inverters with increasing transistor sizes. In [12], a ring oscillator followed by a cascaded inverters output stage delivering –10 dBm over the 3.1–5 GHz range, implemented in 0.18 μm CMOS technology, is presented. This transmitter circuit draws 5 mA from a 1.8 V supply. In [8] a harmonic oscillator followed by a cascaded inverters output stage delivering –10 dBm over the 6.25–8.25 GHz range, implemented in 0.13 μm RFCMOS technology, is presented. This transmitter circuit draws 3.6 mA from a 1.1 V supply.

Figure 4.11 shows a more sophisticated approach exploiting current reuse techniques [13]. The circuit was implemented in a 90 nm CMOS technology.

A differential LC oscillator is implemented by cross-coupled NMOS pair $M_1 - M_2$. The parallel resonant LC tank is formed by the primary winding of transformer T_1 and NMOS varactor CV_1. In order to obtain low supply voltage operation (V_{DD} = 600 mV), transistor gate lengths of 150 nm are used with a threshold voltage of 430 mV. VCO transistors M_1 and M_2 operate at the verge of threshold; VB is chosen to be 380 mV, and VG is 880 mV. Transistors M3 and M4 operate in triode, limiting the isolation of the buffer stage to 10 dB. Supply current is shared by both the oscillator and buffer stages.

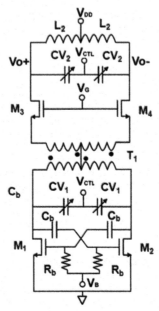

Figure 4.11   LC VCO and output stage with current reuse [13].

This configuration minimizes the power consumption and eliminates the need for a current source. L_2 and CV_2 constitute a parallel resonant LC tank tuned to the FM-UWB center frequency. Inductor L_2 is a small loop antenna. No additional matching network is required, and the VCO chip may be connected via flip chip to the loop antenna to minimize parasitic effects. Its simulated supply current is 700 μA which yields 420 μW of DC power consumption. RF output power is 40 μW (–14 dBm) at 7.75 GHz. Phase noise is –90 dBc/Hz at 1 MHz offset across the band, which is sufficient for the FM-UWB application.

4.3 Conclusions

Direct digital synthesis (DDS) techniques are used for subcarrier generation, yielding a flexible design with low parts count in hardware and accurate output signals in terms of frequency and deviation. Analog and digital techniques are combined for the RF part. The FM-UWB signal is generated by a free-running RF VCO that is calibrated by a PLL frequency synthesizer. VCO tuning curve non-linearity affects the flatness of the power spectral density (PSD) of the FM-UWB signal. Oscillator phase noise limits the performance in the presence of a strong unwanted FM-UWB user in a similar way as multiple-access interference.

When we allow for 1 dB of peaking of the PSD of the FM-UWB signal, the VCO gain at the band edges should not be lower than 80% of the value at the VCO center frequency. It was shown that to be multiple-access limited in a 250 kbps system, the transmitter phase noise needs to be lower than -80 dBc/Hz at 1 MHz offset from the carrier. Such an oscillator can be realized at low resonator power (< 1 μW) when a harmonic oscillator with quality factor of 10 is used. Power consumption of the oscillator will be limited by the parasitic capacitances. Current reuse techniques may be used to stack the oscillator and output amplifier circuit, yielding reduced power consumption.

References

[1] James A. Crawford, *Frequency Synthesizer Design Handbook*, Boston: Artech House, 1994.

[2] Victor S. Reinhardt, "Direct Digital Synthesizers", in *Proc. Seventeenth Annual Precise Time and Time Interval (PTTI) Applications and Planning Meeting*, pp. 345–374.

[3] Peter Nilsson, John F.M. Gerrits, Jiren Yuan, "A Low Complexity DDS IC for FM-UWB Applications," in *Proc. IST Mobile & Wireless Communications Summit 1997*, pp. 1–5.

[4] Private communication from Peter Nilsson, Lund University.

[5] C.J.M Verhoeven, *First order oscillators*, PhD. Thesis, Technische Universiteit Delft, 1989.

[6] Jan R. Westra, Chris J.M. Verhoeven, Arthur H.M. van Roermund, *Oscillators and Oscillator Systems: Classification, Analysis and Synthesis*, New York: Springer US, 2010.

[7] Nelson M. Blachman, George A. McAlpine, "The Spectrum of a High-Index FM Waveform: Woodward's Theorem Revisited," *IEEE Transactions on Communication Technology*, vol. 17, no. 2, pp. 201–208, April 1969.

[8] Marco Detratti, Ernesto Perez, John F.M. Gerrits and Manuel Lobeira, "A 4.2 mW 6.25–8.25 GHz Transmitter IC for FM-UWB Applications," *Proc. ICUWB2009*, pp. 180–184.

[9] Thomas H. Lee, and Ali Hajimiri, "Oscillator Phase Noise: A Tutorial," *IEEE Journal of Solid-State Cicuits*, vol. 35, no. 3, pp. 326–336, March 2000.

[10] P. Andreani, Xiaoyan Wang, L. Vandi, A. Fard, "A study of phase noise in Colpitts and LC-tank CMOS oscillators," *IEEE Journal of Solid-State Circuits*, volume 40, no. 5, pp. 1107–1118, May 2005.

[11] P. Andreani, A. Fard, "More on the $1/f^2$ Phase Noise Performance of CMOS Differential-Pair LC-Tank Oscillators," *IEEE Journal of Solid-State Circuits*, vol. 41, no. 12, pp. 2703–2712, December 2006.

[12] Apostolos Georgiadis, and Marco Detratti, "A Linear, Low-Power, Wide-band CMOS VCO for FM-UWB Applications," *Microwave and Optical Technology Letters*, vol. 50, no. 7, pp. 1955–1958, July 2008.

[13] John Gerrits, Mina Danesh, Yi Zhao, Yunzhi Dong, Gerrit van Veenendaal, John Long, and John Farserotu, "System and Circuit Considerations for Low-Complexity Constant-Envelope FM-UWB," *Proc. ISCAS2010*, pp. 3300–3303.

5

FM-UWB Receiver Implementation

This chapter presents practical implementation examples of the FM-UWB receiver. The envisioned application is body-area networks [1], with a minimum range of 3 meters. Practical circuit implementations are used as vehicles to illustrate the design challenges and trade-offs and as a proof of concept.

Section 5.1 presents the specifications of the receiver based upon both theoretical and practical considerations. Section 5.2 focuses on the wideband FM demodulator. The sensitivity of this block determines the LNA gain that is required to obtain the desired receiver sensitivity. The noise at the multiplier output needs to be lowered without reducing the demodulator gain. Since the large-signal behavior of the front-end is dominated by the FM demodulator, its large-signal performance needs to be optimized. A circuit implementation example of the demodulator exploiting noise reduction and current reuse techniques is presented. Section 5.3 presents a low noise amplifier circuit exploiting noise canceling and current reuse techniques. Section 5.4 presents an implementation example of the FSK subcarrier processor. A classical direct-conversion approach is the starting point. Due to the presence of subcarrier harmonics at the FM demodulator output, additional filtering of the FSK subcarrier signal is required prior to downconversion to baseband. Section 5.5 presents conclusions.

5.1 Receiver Front-end Specification

Figure 5.1 shows the receiver front-end, comprising a bandpass prefilter with gain A_F, LNA with gain A_{LNA}, and the wideband FM demodulator with gain K_D. It is necessary to optimize overall receiver performance in terms of sensitivity and large-signal performance.

The envisioned application is body-area networks, with a minimum range of 3 meters and line-of-sight propagation with multipath as described by the IEEE802.15.6 channel models [2]. The link budget is presented in Table 5.1.

89

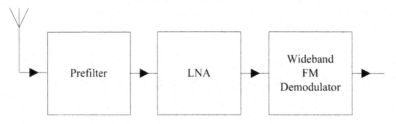

Figure 5.1   FM-UWB receiver front-end.

Table 5.1   FM-UWB link budget for BAN scenario

Parameter	Symbol	Value	Unit	Comments
Tx power	P_{TX}	−14.3	dBm	500 MHz bandwidth
Tx back-off		−2	dB	VCO tuning curve non-linearities
Tx antenna gain	G_{TX}	0.0	dBi	
EIRP	EIRP	−16.3	dBm	P_{TX}-TX back off + G_{TX}
Free space path loss	Lp	−59.5	dB	3 meters range at 7.5 GHz
Rx antenna gain	G_{RX}	0.0	dBi	
RX prefilter	A_F	−2	dB	
Rx power	P_{RX}	**−77.8**	**dBm**	EIRP + Lp + G_{RX} + A_F
Noise power	N_{RX}	−82.0	dBm	500 MHz RF bandwidth, N_{RX} = 5 dB
Required RF SNR	SNR_{RF}	−6.0	dB	$p_b = 1 \times 10^{-6}$, R = 250 kbps
Link margin	M	3.0	dB	Multipath fading, CM3/CM4 channels
RX sensitivity		**−85**	**dBm**	
Remaining margin	M_{rem}	**7.2**	**dB**	Positive margin, link is closed

It is an updated version of the one presented in [3], based upon the theoretical performance (AWGN plus multipath) as presented in Chapter 3, and practical implementation losses due to VCO tuning curve non-linearity as described in Chapter 4 of this book.

Antennas with 0 dBi gain are assumed, and the receiver prefilter is assumed to have 2 dB of insertion loss. The required receiver sensitivity is −85 dBm and the link is closed with 7 dB margin. This margin makes the FM-UWB link operate over double the minimum required distance.

The maximum signal level comes from a FM-UWB transmitter at 50 cm distance without multipath. This yields −58 dBm signal power, and corresponds to a dynamic range of 25–30 dB at the receiver input. Both LNA and FM demodulator need to be able to cope with this range.

5.2 FM Demodulator Implementation

The key receiver building block is the wideband FM demodulator that is not preceded by a hard-limiting device. As a result, the FM capture effect [4] doesn't occur. This allows for the simultaneous demodulation (despreading) of multiple FM-UWB input signals at the same center frequency with different subcarrier frequencies. It was shown in Section 2.3.1 that the output voltage of an ideal FM demodulator based upon a multiplier with lossless delay line, with delay time $\tau = N/4f_C$ and input signal amplitude A is given by

$$V_{DEM}(f) = \frac{A^2}{2}\cos\left(\frac{N\pi f}{2f_C}\right) = -\frac{A^2}{2}\sin\left(N\frac{\pi}{2}\left[\frac{f}{f_C} - 1\right]\right). \quad (5.1)$$

Figure 5.2 shows the block diagram of a practical implementation of the fixed time delay FM demodulator (compared to the ideal demodulator of Figure 2.8). The FM signal V_{RF} is transformed into a PM signal (V_Y) by the delay circuit with transfer function $H(j\omega)$ and delay time τ. Direct and delayed signals are multiplied to obtain the demodulated signal V_{DEM}. Voltage gain may be present in both the direct and delayed paths (A_{VDEM1} and A_{VDEM2}, respectively) Total voltage gain $A_{VDEMTOT}$ is defined as

$$A_{VDEMTOT} = A_{VDEM1}\text{x}A_{VDEM2}. \quad (5.2)$$

The gain distribution factor k_g is defined as

$$k_g = \frac{A_{VDEM2}}{A_{VDEM1}}. \quad (5.3)$$

The individual gains are given by

$$A_{VDEM1} = \sqrt{\frac{A_{VDEMTOT}}{k_g}}, \quad (5.4)$$

$$\text{and } A_{VDEM2} = \sqrt{k_g A_{VDEMTOT}}. \quad (5.5)$$

The non-linearities of the multiplier are modeled by blocks NL_1 and NL_2, with small-signal gains of unity.

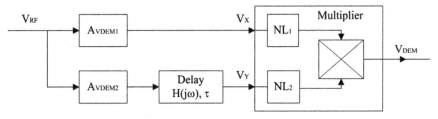

Figure 5.2 Fixed time delay FM demodulator with delay circuit and gain in both branches.

We consider the case of a single-frequency sinusoidal input signal. Let V_X and V_Y be the two multiplier input signals with amplitudes A_X and A_Y and phases 0 and ϕ_Y respectively, such that

$$V_X = A_{VDEM1}V_{RF} = A_X \cos(\omega t),$$
$$\text{and } V_Y = A_{VDEM2}H(V_{RF}) = A_Y \cos(\omega t + \phi_Y). \tag{5.6}$$

When the delay circuit is implemented as an ideal, (i.e., lossless) delay line, both A_X and A_Y are independent of frequency and the phase difference between signals V_X and V_Y (i.e., $\phi = \phi_X - \phi_Y$) equals

$$\phi(f) = -2\pi f \tau. \tag{5.7}$$

As shown in Section 2.3.1, optimum performance implies maximum amplitude of the fundamental of the subcarrier signal in the demodulator output signal. The phase lag at the center frequency needs to be

$$\phi(f_C) = -2\pi f_C \tau = -2\pi f_C N \frac{T}{4} = -N \frac{\pi}{2}. \tag{5.8}$$

This requires a delay time (τ) equal to an odd multiple of a quarter period (T/4) at the center frequency f_C of the FM signal. The delay time of an ideal delay line is constant and equal to the group delay time τ_g (i.e., the derivative of the phase response).

$$\tau = -\frac{\partial \phi}{\partial \omega} = N\frac{T}{4} = \frac{N}{4f_C}, \quad \text{for } N = 1, 3, 5 \dots \tag{5.9}$$

The FM demodulator bandwidth (B_{DEM}) is given by

$$B_{DEM} = \frac{1}{2\tau} = \frac{2}{N}f_C. \tag{5.10}$$

It can't be chosen freely, since N must be an odd integer number. For example, for a center frequency of 7.5 GHz and 500 MHz bandwidth, N should be either 29 or 31, which changes the demodulator bandwidth by only 6% (i.e., 30 MHz).

Figure 5.3 shows amplitude and phase response (modulo 2π) for a FM demodulator using a delay line. The center frequency is 7.5 GHz, and demodulator bandwidth of approximately 500 MHz is obtained by choosing N = 31. Figure 5.4a shows $|H(f)|$ and $\cos(\phi(f))$ versus frequency for this delay line, and Figure 5.4b shows the triangular subcarrier signal and the demodulator output voltage in the absence of multipath. The output signal, $V_{DEM}(t)$, is a sine wave without distortion.

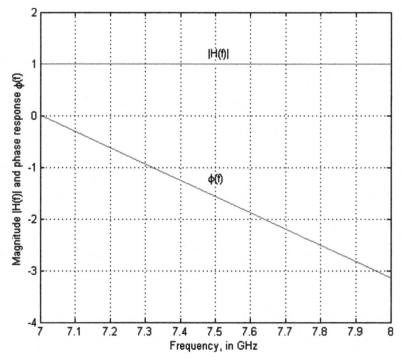

Figure 5.3 Magnitude and phase response of the ideal delay line used in a FM demodulator with $f_C = 7.5$ GHz and $B_{DEM} = 500$ MHz.

5.2.1 Delay Circuit Specification and Implementation

Continuous tuning of the FM demodulator parameters is a necessary requirement for a flexible FM-UWB receiver that can operate at multiple frequencies and various RF bandwidths (e.g., 500 MHz and 1 GHz).

The required delay time, τ, depends on the demodulator bandwidth and follows from (5.10). Some fine tuning is required to satisfy the phase lag condition given by (5.9). To cover a frequency range [f_{CMIN}, f_{CMAX}], the required ratio of maximum and minimum delay time is given by

$$\frac{\tau_{MAX}}{\tau_{MIN}} = \frac{1}{2}\left(1 + \frac{f_{CMAX}}{f_{CMIN}}\right) \tag{5.11}$$

This means, e.g., that tuning the center frequency over the 7–8 GHz band requires a ratio of 1.07. Changing the bandwidth requires a change in the delay time and thus, the physical length of the delay line.

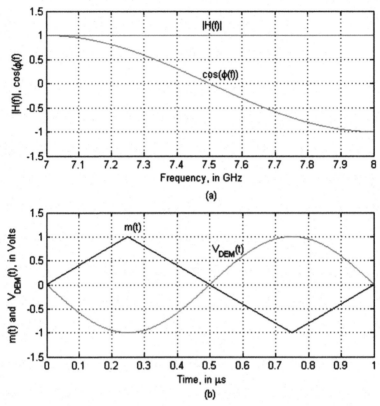

Figure 5.4   (a) |H(f)| and cos(φ(f)), (b) transmitted subcarrier signal and demodulator output voltage for the delay line demodulator with f_C = 7.5 GHz and B_{DEM} = 500 MHz.

Realization of a true, tunable time delay typically requires greater chip area [5, 6] and additional power consumption for the driving electronics. Such a delay line has a flat amplitude response and provides no rejection of signals in adjacent RF channels.

When the time delay is implemented by the group delay of a resonator, both demodulator center frequency and bandwidth can be tuned continuously, and over a wide range without changing the physical size of the components. Moreover, a resonator provides frequency selectivity, which is advantageous for attenuating signals in adjacent RF channels.

Referring to Figure 5.2, replacing the lossless delay line by a resonator circuit implies that the amplitude and phase of the delayed signal become frequency dependent. The resonator quality factor determines the slope of the phase response, but does not influence the value of the center frequency. The demodulator output signal versus frequency is given by

$$V_{DEM}(f) = K_D A_X A_Y(f) \cos(\phi_Y(f))$$
$$= K_D A_{VDEMTOT} A_{RF}^2 |H(f)| \cos\phi(f) \qquad (5.12)$$

The required delay time at the demodulator center frequency, i.e., the filter group delay yielding the same output signal as a delay line with delay time τ, is given by (5.9). A continuous range of demodulator bandwidth values can be implemented by changing the resonator quality factor (Q) as the resonator group delay is proportional to Q. Optimum performance is obtained when the phase lag of the delay circuit is consistent with (5.9).

In the time domain, the instantaneous amplitude, $A_Y(t)$, and phase $\phi_Y(t)$ of V_Y vary with the instantaneous frequency

$$f_{INST}(t) = f_C + \Delta f x m(t). \qquad (5.13)$$

From Equations (5.12) and (5.13), the demodulator output voltage is therefore given by

$$V_{DEM}(t) = K_D A_{VDEMTOT} V_{RF}^2 |H(f_{INST}(t))| \cos(\phi(f_{INST}(t))). \qquad (5.14)$$

As mentioned in Section 3.3, frequency-selective multipath creates harmonic distortion in the demodulator output signal. In a similar way, amplitude and phase variations (over frequency) in the delay circuit yield harmonic distortion. The difference is that this phenomenon only occurs in one branch of the FM demodulator. Whereas the phase of the FM-UWB signal prior to demodulation is irrelevant, the phase versus frequency dependence of the delay circuit is essential for optimum FM demodulator performance.

When (5.9) is not satisfied, the demodulator transfer function will not be zero at the center frequency, which corresponds to an offset between demodulator center frequency and received FM-UWB signal. In Section 2.3.1.1, it was shown that such a frequency offset lowers the level of the fundamental of the subcarrier signal in the demodulator output, and that even-order harmonic distortion will occur. The amplitude of the fundamental (at f_{SUB}) in the demodulator output is straightforward to calculate. We consider the case where the resonator transfer function H(f) has even symmetry in the magnitude response and odd symmetry in the phase response around the center frequency f_C. This is the case for most resonator circuits with complex conjugate poles and real zeroes.

$$|H(f_C + \delta f)| = |H(f_C - \delta f)|$$

$$\phi(f_C + \delta f) = -\pi - \phi(f_C - \delta f), \tag{5.15}$$

where δf is the offset from the center frequency. The Fourier series of the demodulator output signal is composed of sine wave components only, and the fundamental amplitude (HD_1) is given by

$$HD_1 = \frac{2}{T_{SUB}} \int_0^{T_{SUB}} |H(f_{INST}(t))| \cos(\phi(f_{INST}(t))) \sin(\omega_m t) dt$$

or

$$HD_1 = \frac{4}{T_{SUB}} \int_{-\frac{T_{SUB}}{4}}^{\frac{T_{SUB}}{4}} |H(f_{INST}(t))| \cos(\phi(f_{INST}(t))) \sin(\omega_m t) dt, \tag{5.16}$$

with $T_{SUB} = 1/f_{SUB}$. Due to the triangular waveform, the following substitutions can be made over the integration interval $[-\frac{1}{4}T_{SUB}, \frac{1}{4}T_{SUB}]$, where the instantaneous frequency linearly increases from $f_C - \Delta f$ to $f_C + \Delta f$,

i. $H(f_{INST}(t))$ is replaced by $H(f)$ with $f_C - \Delta f < f < f_C + \Delta f$,
ii. $\cos(\phi(f_{INST}(t)))$ is replaced by $\cos(\phi(f))$ with $f_C - \Delta f < f < f_C + \Delta f$,
iii. $\sin(\omega_m t)$ is replaced by $\sin\left(\frac{\pi}{2}\frac{f - f_C}{\Delta f}\right)$,
iv. $\frac{dt}{T_{SUB}}$ is replaced by $\frac{df}{4\Delta f}$.

The integral can now be calculated with respect to the frequency variable, f, and this yields

$$HD_1 = \frac{2}{2\Delta f} \int_{f_C - \Delta f}^{f_C + \Delta f} |H(f)| \cos(\phi(f)) \sin\left(\frac{\pi}{2}\frac{f - f_C}{\Delta f}\right) df. \tag{5.17}$$

It can be shown that the highest subcarrier output signal is obtained for a delay circuit with a flat amplitude response and a phase response that decreases linearly from 0 to $-\pi$ over the frequency interval $[f_C - \Delta f, f_C + \Delta f]$. Such a circuit behaves similar to an ideal delay line over the range $-\pi < \arg(H(f)) < 0$ and satisfies the phase requirement from (5.9).

In our quest for a resonator circuit that yields optimum performance, Section 5.1.1 investigates the simplest circuit implementation, (i.e., the parallel resonant circuit), and verifies whether its performance is sufficient. Section 5.1.1.2 addresses a slightly more advanced circuit and Section 5.1.1.3 presents a delay circuit implementation that meets all the requirements. An example case for a 7.5 GHz FM demodulator is presented.

Subsection 5.1.2 takes a closer look at the multiplier implementation. Many analog multiplier circuits are known [7], but not all of them suited for the FM-UWB radio. Section 5.1.3 addresses noise originating from the multiplier circuit and ways to lower it. Section 5.1.4 addresses the effect of non-linearities in the receiver front-end and the optimum gain distribution inside the FM demodulator.

5.2.1.1 Parallel resonant circuit as a time delay

Figure 5.5 shows a parallel resonant circuit with inductance L, capacitance C and equivalent parallel loss resistance R_p. This one-port resonator acts as a bandpass filter (BPF) transforming input current I_I into output voltage V_O.

The impedance Z in the frequency (s) domain, resonant frequency ω_0 and quality factor Q of this circuit are given by

$$Z = \frac{V_O}{I_I} = R_p \frac{\frac{\omega_0}{Q}s}{s^2 + \frac{\omega_0}{Q}s + \omega_0^2}, \tag{5.18}$$

$$\omega_0 = \frac{1}{\sqrt{LC}}, \tag{5.19}$$

$$Q = \frac{R_P}{\omega_0 L} = R_p \sqrt{\frac{C}{L}}. \tag{5.20}$$

The phase response of the resonator is relatively linear over the bandwidth f_C/Q, where the magnitude response drops by 3 dB at the band edges. The resonator quality factor can be tuned without affecting the resonator center frequency by adjusting the resistance R_p. The peak value of the resonator group delay τ_{gmax} occurs for $\omega = \omega_0$ and equals

$$\tau_{g\,max} = -\left[\frac{\partial}{\partial\omega}\phi(Z)\right]_{\omega=\omega_0} = \frac{2Q}{\omega_0}. \tag{5.21}$$

The parallel resonant circuit can be used as the load of a gain stage, e.g., a differential pair, and several resonant gain stages can be cascaded to obtain higher delay values. However, since $\phi(f_C)$ equals 0, it is necessary to generate an additional phase shift of $-90°$ or an odd multiple of this value, in order to use this resonator in the constant delay FM demodulator. The additional phase shift may be approximated with a capacitively-degenerated differential

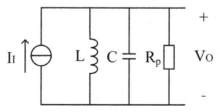

Figure 5.5   Parallel resonant circuit.

pair [8]. This approach can't yield the required constant −90° phase shift, and therefore results in a design compromise with sub-optimum performance.

5.2.1.2 Lattice bandpass filter circuit as a time delay

By using a different resonator topology which realizes a $\phi(f_C)$ of −90°, the problem encountered with the parallel resonant circuit is solved. The lattice bandpass filter (LBPF) resonator shown in Figure 5.6 is the current-driven version of the lattice allpass filter (APF) [9]. Normally this circuit is voltage driven and yields a flat amplitude response, and a phase response that varies from 0 degrees at low frequencies to −180 degrees for frequencies (ω) much greater than ω_0.

When the circuit is current-driven, its transimpedance (V_O/I_I) strongly resembles the impedance of the parallel resonant circuit. Due the presence of an additional zero, the phase shift at the center frequency $\phi(f_0)$ now equals −90 degrees. The transimpedance (Z) of the LBPF in the frequency (s) domain is given by

$$Z = \frac{V_O}{I_I} = -Z_0 \frac{(s + \omega_0)(s - \omega_0)}{s^2 + \frac{\omega_0}{Q}s + \omega_0^2} \qquad (5.22)$$

$$\text{where } Z_0 = \sqrt{\frac{L}{C}} = \omega_0\, L, \qquad (5.23)$$

$$\omega_0 = \frac{1}{\sqrt{LC}}, \qquad (5.24)$$

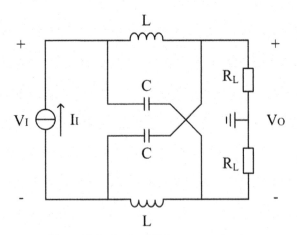

Figure 5.6   Lattice BPF resonator circuit.

$$\text{and } Q = \frac{\omega_O L}{4 R_L} = \frac{Z_O}{4 R_L} \tag{5.25}$$

The LBPF behaves like a lumped-element version of $\lambda/4$ transmission line. Its trans-impedance $Z(\omega_0)$ is independent of the value of termination resistor R_L. The input impedance (Z_I) at resonant frequency ω_0 can be shown to be equal to

$$Z_I(\omega_0) = \frac{V_I}{I_I} = \frac{Z_0^2}{R_L}. \tag{5.26}$$

The maximum value of the group delay occurs at the resonant frequency and equals

$$\tau_{g\,\max} = \frac{2Q}{\omega_0} = \frac{L}{2 R_L}. \tag{5.27}$$

Figure 5.7 shows the normalized magnitude response, as well as the phase response and the group delay of the transfer function (V_O/I_I) of the parallel resonant BPF circuit and the lattice BPF resonator circuit for a resonant frequency of 1 rad/s and quality factor equal to 5.

The phase response of the LBPF resonator is relatively linear over a bandwidth f_C/Q, and the magnitude response drops by 3 dB at the band edges as shown in Figure 5.7a. The advantage of the LBPF is the $-90°$ phase shift at the resonant frequency as seen in Figure 5.7b. A disadvantage is the lower gain, which is a factor Q lower than the parallel resonant circuit. At equal gain, this means higher current consumption.

5.2.1.3 Cascade of BPF and LBPF as a time delay

The best of both worlds, i.e., $-90°$ phase shift at the resonant frequency and reasonable gain is obtained by cascading one or several BPF resonators and a LBPF section. Assuming no interaction between the resonators, the magnitude and phase responses, the group delay and equivalent quality factor of the overall transfer function are given by

$$A_{TOT}(\omega) = A_1(\omega) A_2(\omega) \tag{5.28}$$
$$\phi_{TOT} = \phi_1 + \phi_2 \tag{5.29}$$
$$\tau_{TOT} = \tau_1 + \tau_2 \tag{5.30}$$
$$Q_{TOT} = Q_1 + Q_2 \tag{5.31}$$

The overall group delay is the sum of the individual group delays (5.30) and the phase shift of $-90°$ is taken care of by the LBPF. This is illustrated in Figure 5.8 for the case where 80% of the delay time is realized in the BPF. Normalization

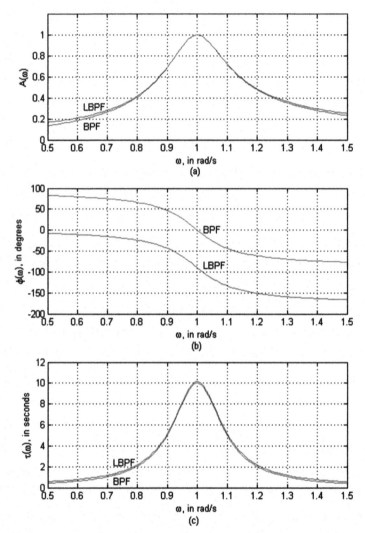

Figure 5.7   Magnitude (normalized to one), phase and group delay of parallel resonant circuit (BPF) and lattice resonant circuit (LBPF) with Q = 5 and ω_0 = 1.

of the group delay is with respect to the delay time of a resonator with a Q of unity, i.e., $\tau = 2/\omega_0$.

The cascaded resonator approach has an additional advantage. A single resonator cannot cover the [0 – 180°] range with its phase. A cascade of two resonators can.

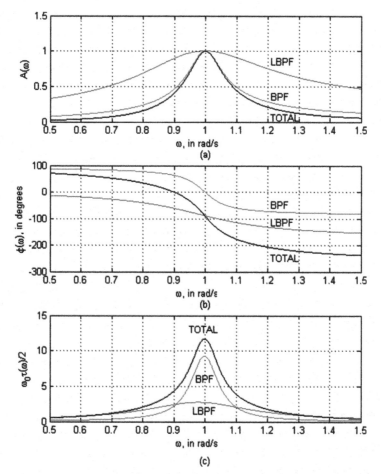

Figure 5.8 Magnitude (normalized to one), phase and normalized group delay for the cascaded BPF-LBPF delay circuit.

Practical Example

Figure 5.9 shows the block diagram of a wideband FM demodulator based upon the cascade of two gain stages driving a LBPF and BPF as described by Dong in [10].

The circuit operates at a center frequency of 7.45 GHz and has a bandwidth of 1 GHz, which would require a delay time of 505 ps (5.9) when realized with a true delay line. Both gain stages have a transconductance of 14 mA/V. The LBPF stage has a transimpedance, Z_0, of 120 Ω and the load impedance, Rp, of the BPF is 900 Ω. Overall gain of the two stages is 26 dB, most of which

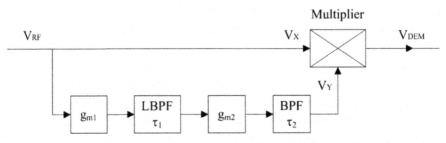

Figure 5.9   FM demodulator using a cascade of LBPF and BPF as delay circuit.

is realized in the gain stage by a parallel resonant circuit. The impedance of the parallel resonant circuit is proportional to the quality factor. The high gain (g_mR_p) of the BPF stage requires a high quality factor ($Q_{BPF} = 9.3$), and as a result 80% of the delay is realized in this stage. The remainder of the delay and the $-90°$ phase shift is realized in a LBPF section with lower quality factor ($Q_{LBPF} = 2.5$) as shown in Table 5.2.

The detailed circuit schematics can be found in Appendix C. The LBPF and BPF stages are active; a differential pair amplifier drives each resonator.

Figure 5.10 shows the magnitude and phase responses of the delay circuit, $|H(f)|$ and $\phi(f)$, respectively. Figure 5.11 shows the triangular subcarrier signal and the demodulator output voltage, $V_{DEM}(t)$, which is now a distorted sine wave.

Figure 5.12 shows how the level of the fundamental of the subcarrier signal at the FM demodulator output varies with the quality factor (Q_{TOT}) of the delay circuit. The reference level of 0 dB corresponds to the value obtained with a true delay line. The signal level increases until the delay time corresponding to (5.9) has been achieved, i.e., a quality factor of 12, and next flattens out up to the double of the optimum Q value. HD_1 remains constant since the output signal resembles a square wave. There is a broad peak for Q between 10 and 20. The subcarrier level is 3.3 dB lower than an ideal delay line demodulator. This is due to the fact that the magnitude response of the delay circuit is not flat, as can be seen in Figure 5.11. Since the delay circuit also includes gain stages, this loss can be easily compensated for.

Table 5.2   Partitioning of delay time in practical demodulator circuit [5]

Delay Circuit	Q	τ, in ps	A_V	A_V in dB
LBPF	2.5	107	1.6	4
BPF	9.3	398	12.6	22
TOTAL	11.8	505	20	26

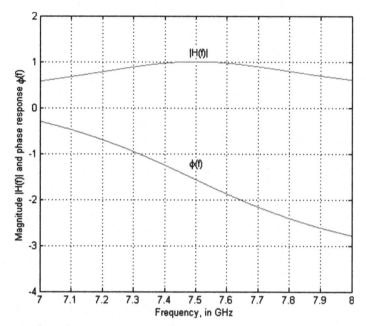

Figure 5.10 |H(f)| and cos(ϕ(f)) of the LBPF-BPF cascade with Q_{TOT} = 12 used in a FM demodulator with f_C = 7.5 GHz.

The gain (A_{VDEM}) in the delay circuit improves the demodulator sensitivity (at RF) by a factor $\sqrt{A_{VDEM}}$, or one half of its voltage gain in decibels, i.e., 13 dB. The bias current for these amplifiers is reused from the multiplier circuit in order to reduce the overall power consumption. However, it will be shown in Section 5.1.4 that adding gain in one branch of the FM demodulator is not optimum for the large-signal behavior.

5.2.2 Multiplier Implementation

In the wideband FM demodulator of the FM-UWB radio, it is important to use a multiplier circuit that yields high gain, modest linearity and low power consumption. The double-balanced Gilbert multiplier as shown in Figure 5.13 meets these requirements [11, 12].

The output voltage of this multiplier circuit biased at a current I_T is given by

$$V_O = I_T R_C \tanh\left(\frac{V_X}{2V_T}\right) \tanh\left(\frac{V_Y}{2V_T}\right) \approx 400 I_T R_C V_X V_Y, \qquad (5.32)$$

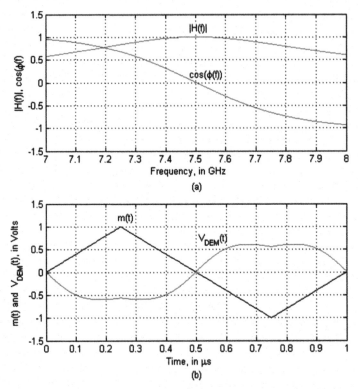

Figure 5.11   (a) Magnitude and phase responses, (b) transmitted subcarrier signal and demodulator output voltage for the LBPF demodulator with $f_C = 7.5$ GHz and $Q_{TOT} = 12$.

$$\text{where } V_T = \frac{kT}{q} \approx 25\text{mV}.$$

The approximation given in (5.32) is valid for input voltages V_X, V_Y much less than V_T. The term $I_T R_C$ is equal to the maximum voltage drop across either R_{C1} or R_{C2}. For small input signal amplitudes (i.e., V_X, $V_Y \ll V_T$), the voltage drop across both R_{C1} and R_{C2} equals ½ $I_T R_C$. A typical value for $I_T R_C$ in a practical circuit is 500 mV.

In the FM demodulator application, one of the input signals is a delayed version of the other one, i.e.,

$$V_Y(t) = V_X(t - \tau). \tag{5.33}$$

When the delay time $\tau = 0$, both V_X and V_Y equal V_{RF}, and the output signal V_O equals

$$V_O = K_D V_{RF}^2 \tag{5.34}$$

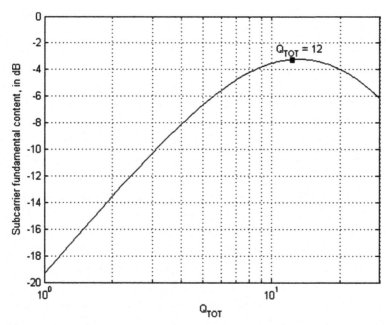

Figure 5.12 Subcarrier fundamental level versus the quality factor of the LBPF-BPF delay circuit.

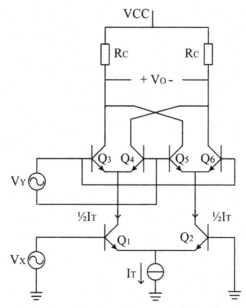

Figure 5.13 Double-balanced Gilbert multiplier.

For the double-balanced mixer (DBM) the demodulator sensitivity, K_D, equals

$$K_D = \frac{I_T R_C}{4V_T^2} \approx 400 I_T R_C. \tag{5.35}$$

From (5.35) it follows that for $I_T R_C$ of 0.5 V and K_D of 200 V^{-1} a sinusoidal input signal with amplitude (A_{RF}) of 10 mV will yield an output signal with a DC component of 10 mV. This is also the peak value (A_{SUB}) of the demodulated signal that is obtained in a FM demodulator circuit with an overdrive equal to one (see Section 2.3.1 for the overdrive definition). The amplitude A_{SUB} of the lowpass-filtered multiplier output signal (i.e., a sine wave at the subcarrier frequency f_{SUB}) is related to the amplitude A_{RF} of the FM-UWB signal by

$$A_{\text{SUB}} = \frac{1}{2} K_D A_{RF}^2 = \frac{I_T R_C}{8 V_T^2} A_{RF}^2. \tag{5.36}$$

For an overdrive of 1 and demodulator sensitivity of 200 V^{-1}, the input and output signal amplitudes are the same for an input signal amplitude of 10 mV (−30 dBm in a 50 Ω system). Figure 5.14 shows the FM-UWB input signal, FM-UWB frequency spectrum, and demodulator output signal for a center frequency of 7.5 GHz and 500 MHz bandwidth.

The demodulator sensitivity of the double-balanced Gilbert multiplier with additional preamplification (see Figure 5.2), is given by

$$K_D = A_{VDEM1} A_{VDEM2} \frac{I_T R_C}{4 V_T^2}. \tag{5.37}$$

5.2.3 FM Demodulator Noise

This subsection analyzes the noise originating from the active parts of the multiplier circuit and the effect on the FM demodulator and the overall receiver front-end performance. The analysis presented here is similar to the one in Chapter 3 (see Figure 3.2), but takes also into account noise from the multiplier circuit. Figure 5.15 shows a block diagram of the receiver front-end with the equivalent noise source V_N at the receiver input and equivalent output noise voltage source V_{NO} that models the low-frequency noise (at f_{SUB}) from the multiplier electronics. Small-signal conditions are assumed. Sources V_N and V_{NO} are uncorrelated white noise sources. The signal path becomes differential after the LNA.

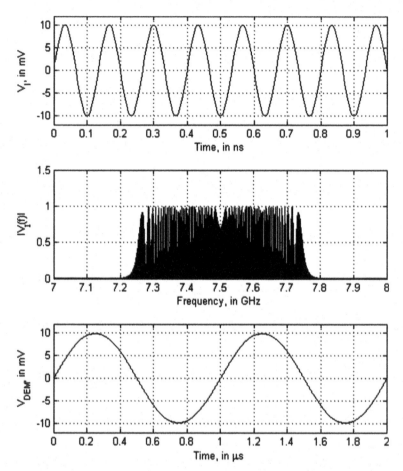

Figure 5.14 FM-UWB signal with f_C = 7.5 GHz, B_{RF} = 500 MHz and A_{RF} = 10 mV, its normalized spectrum and the FM demodulator output signal for K_D = 200 V^{-1} and $I_T R_C$ = 0.5 V.

Source V_N represents the thermal noise floor of the RF input signal V_S plus the equivalent input noise of the LNA. The RMS value of the input noise is equal to

$$V_N = \sqrt{4kTRB_{RF}F_{LNA}},\qquad(5.38)$$

where F_{LNA} is the noise factor of the LNA for a 50 Ω source in a 1 Hz bandwidth.

The demodulator input signal (V) is given by

$$V_{RF} = A_V(V_S + V_N).\qquad(5.39)$$

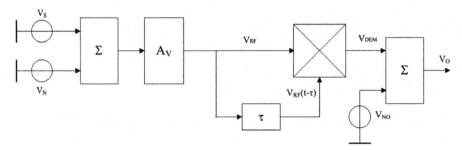

Figure 5.15   Overall receiver frontend with noise sources.

The demodulator output voltage is given by

$$V_O = V_{DEM} + V_{NO} = K_D V_{RF}^2 + V_{NO}. \tag{5.40}$$

(5.40) is the practical version of (3.4). We further define the function T(f) as

$$T(f) = K_D A_V^2(f) \tag{5.41}$$

We define signal power S, equivalent input noise power N and equivalent output noise power N_{NO} as

$$S = V_S^2$$
$$N = V_N^2 \tag{5.42}$$
$$N_{NO} = V_{NO}^2,$$

These three quantities are uncorrelated. Solving this in a manner similar to that shown in Section 3.1 yields the output signal power (S_O) and the total output noise power (N_O) in the subcarrier bandwidth

$$S_O = V_{DEM}^2 = |T(f)|^2 V_S^4 = |T(f)|^2 S^2, \tag{5.43}$$

$$N_O = \frac{B_{SUB}}{B_{RF}} |T(f)|^2 (N^2 + 4SN) + N_{NO}$$

$$= |T(f)|^2 \left(\frac{B_{SUB}}{B_{RF}} [N^2 + 4SN] + \frac{N_{NO}}{|T(f)|^2} \right) \tag{5.44}$$

The 4SN term in (5.44) originates from the non-linearity of the multiplication and was already encountered in (3.4).

5.2.3.1 Noise from the FM demodulator with Gilbert multiplier

The noise at the output of the Gilbert multiplier circuit (Figure 5.13) is mainly constituted by the collector shot noise and thermal noise of the base resistance

(R_B) of transistors Q_{3-6} [13]. We define the quad current (I_{quad}) as the sum of the collector currents of transistors Q_{3-6}.

$$I_{quad} = \sum_{i=1}^{4} I_{Ci} \tag{5.45}$$

In the standard version of the Gilbert multiplier (Figure 5.13) the quad current is equal to the tail current (I_T). For the Gilbert multiplier shown in Figure 5.13 the mean squared output noise voltage per Hz (expressed in V^2/Hz) due to the collector shot noise of the quad transistors is equal to

$$\overline{V}^2_{NO,shot} = 4qI_{quad}R_C^2. \tag{5.46}$$

The voltage spectral density due to the thermal noise of the base resistance (R_B) of the quad transistors is

$$\overline{V}^2_{NO,RB} = 64kTR_Bg_m^2R_C^2, \tag{5.47}$$

with g_m the transconductance of a differential pair biased at ½I_{quad}

$$g_m = \frac{I_{quad}}{8V_T} \tag{5.48}$$

Substituting (5.48) in (5.47) yields the mean squared output noise voltage per Hz due to the base resistance thermal noise of the quad transistors

$$\overline{V}^2_{NO,RB} = \frac{I_{quad}R_B}{V_T}qI_{quad}R_C^2. \tag{5.49}$$

The total noise power (P_{NO}) in the subcarrier bandwidth (B_{SUB}) of the quad noise at the Gilbert multiplier output therefore equals

$$N_{NO} = 4qI_{quad}R_C^2\left(1 + \frac{I_{quad}R_B}{4V_T}\right)B_{SUB}. \tag{5.50}$$

The RMS value V_{NO} of the output noise voltage in the subcarrier bandwidth B_{SUB} is equal to

$$V_{NO} = 2R_C\sqrt{qI_{quad}\left(1 + \frac{I_{quad}R_B}{4V_T}\right)B_{SUB}} \tag{5.51}$$

It can be seen that the collector shot noise dominates when the quad current $I_{quad} \ll 4V_T/R_B$. Downconverted noise from Q_1 and Q_2 can be ignored, as long as the conversion gain of the multiplier quad is low, i.e., as long as the amplitude of the input signal is much smaller than V_T.

An Example

To get a better feeling for the numbers, an example is given here for a 250 kbps system with an RF bandwidth of 500 MHz and subcarrier bandwidth of 500 kHz. The Gilbert multiplier is biased at a current ($I_T = I_{quad}$) of 1 mA and collector resistors R_C have a value of 500 Ω. The quad transistors are assumed to have a base resistance of 50 Ω.

This yields an output noise voltage of 11 μV. For a probability of error of 1×10^{-6}, a subcarrier SNR of 14 dB is required, corresponding to a demodulator output signal of 55 μV_{RMS}. Using (5.36), it follows that a demodulator input signal amplitude of 764 μV_{RMS} (–49.3 dBm in a 50 Ω system) is necessary. This is the sensitivity of the example FM-UWB receiver without an LNA.

This compares unfavorably to the thermal noise power in a 500 MHz bandwidth, which is equal to –87 dBm. A LNA with a gain higher than 30 dB is therefore required in order to obtain a receiver sensitivity close to the noise floor. Lowering the noise of the multiplier is therefore worthwhile, because then the LNA gain can be lowered, yielding lower power consumption, smaller chip area and improved linearity performance, since the FM demodulator is driven with a smaller signal.

When the receiver is operating at its theoretical sensitivity limit, the SNR at the receiver input is smaller than unity and the 4SN term in (5.44) can be neglected. In order to obtain this sensitivity level, it is required that the FM demodulator output noise is dominated by the amplified RF noise and not the demodulator output noise. (5.44) then transforms into the following inequality

$$\frac{B_{SUB}}{B_{RF}}|T(f)|^2 N^2 > N_{NO}. \tag{5.52}$$

Substituting (5.41) for T(f) and solving for A_V yields

$$A_V^4 > \frac{N_{NO}}{K_D^2 N^2} \frac{B_{RF}}{B_{SUB}}. \tag{5.53}$$

Substituting (5.50) for N_{NO} and (5.38) and (5.42) for N yields

$$A_V^4 > \frac{4qI_{quad}R_C^2}{\left(16k^2T^2R^2B_{RF}^2F_{LNA}^2\right)K_D^2}\left(1 + \frac{I_{quad}R_B}{4V_T}\right)B_{RF}. \tag{5.54}$$

Substituting (5.37) with $A_{VDEM1} = A_{VDEM2} = 1$ for K_D yields

$$A_V^4 > \frac{4qI_{quad}R_C^2}{\left(16k^2T^2R^2B_{RF}^2F_{LNA}^2\right)}\frac{16V_T^4}{I_T^2R_C^2}\left(1 + \frac{I_{quad}R_B}{4V_T}\right)B_{RF}. \tag{5.55}$$

Substituting kT/q for V_T yields

$$A_V^4 > \frac{4qI_{quad}R_C^2}{\left(16k^2T^2R^2B_{RF}^2F_{LNA}^2\right)}\frac{16k^4T^4}{q^4I_T^2R_C^2}\left(1 + \frac{I_{quad}R_B}{4V_T}\right)B_{RF}, \quad (5.56)$$

which can be simplified to

$$A_V^4 > \frac{4k^2T^2}{q^3R^2B_{RF}F_{LNA}^2}\frac{I_{quad}}{I_T^2}\left(1 + \frac{I_{quad}R_B}{4V_T}\right) \quad (5.57)$$

It can be seen directly from (5.57) that to lower the LNA gain, it is good to have a high current (I_T) in the differential pair Q_{1-2} of the Gilbert multiplier, and a low current in the quad transistors Q_{quad} is desired. The following subsection addresses this current bleeding approach.

5.2.3.2 Reduction of the Gilbert multiplier noise by current bleeding

In order to lower the noise of the multiplier circuit of the FM demodulator, one can lower the bias current of transistors Q_{3-6} in Figure 5.13. This can be done by diverting current from the collectors of transistors Q_1 and Q_2. There is no change in the demodulator gain (K_D) if the transconductance of transistors Q_1 and Q_2 remains unaltered. (5.32) remains valid if the quad current I_{quad} is smaller than the mixer tail current I_T.

The diverted current is not necessarily lost, but can be reused to bias the delay circuit that drives one input of the multiplier. Figure 5.16 shows circuit details of a wideband demodulator using this principle [10]. Complete schematics can be found in Appendix C.

The current in the quad transistors (I_{quad}) is reduced by a factor of 10 and the multiplier noise is lowered by 10 dB. This yields a 5 dB demodulator sensitivity increase, meaning that 5 dB lower LNA gain is required for the same performance. This is illustrated in curves 1 and 2 of Figure 5.17, which shows the receiver sensitivity, i.e., the RF input level that yields a subcarrier SNR of 14 dB versus LNA gain for a data rate of 250 kbps. The target receiver sensitivity for the BAN applications is −85 dBm. The first curve shows the starting point, which is maximum current in the quad. The second curve shows the effect of lowering of the quad current. The diverted current is reused in the gain stages of the delay circuit and yields an additional gain of 26 dB in the delayed path. As a result the demodulator sensitivity is increased by another 13 dB. This is shown in the third curve.

Increasing the delay path gain is one option to increase the demodulator sensitivity, however, the large signal behavior will be degraded. Moreover, gain (A_{VDEM2}) created in a single branch will only increase the demodulator

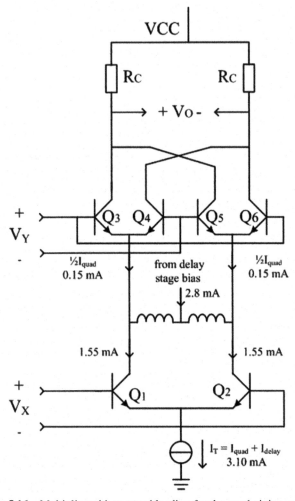

Figure 5.16   Multiplier with current bleeding for the quad giving reduced.

sensitivity by its square root ($\sqrt{A_{VDEM2}}$). It is therefore better to create additional gain in front of the FM demodulator in order to increase receiver sensitivity.

It will be shown in the next section that the best overall receiver performance is obtained by having unity gain in both the direct and delayed branches. Part of the diverted current is required for this purpose. The remaining current should be used to bias an intermediate gain stage between LNA and FM demodulator. This compromise increases the receiver sensitivity without degrading the large-signal behavior.

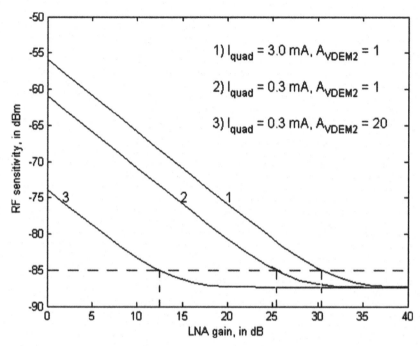

Figure 5.17 Receiver sensitivity versus LNA gain for various values of I_{quad} and with additional gain in the delayed branch.

5.2.4 Implications of Non-linearities in the FM-UWB Receiver Front-end

Circuit non-linearities determine the large-signal performance of radio receiver circuits. They set a limit on the largest signal that can be processed and may compromise the reception of a weak signal when multiple input signals are present [14]. Non-linearities cause harmonic distortion, intermodulation (IM) distortion, and gain compression.

The transfer function y(x) of a non-linear device may be modeled by a Taylor series of the form

$$y(x) = k_0 + k_1 x + k_2 x^2 + k_3 x^3 + \cdots + k_n x^n. \tag{5.58}$$

The k_0 term represents the DC offset at the output in the absence of an input signal. In an RF system with bandpass transfer and AC coupling, this DC offset is removed. The k_1 term represents the linear gain of the device and is not responsible for any distortion. The higher order terms result in an output signal that contains components that don't exist in the input signal. When a single frequency input at f_1 is applied to the input of a non-linear device, its output will contain harmonics at frequencies nf_1. When two frequencies f_1 and f_2 are applied to a non-linear device, its output signal contains components at frequencies $\pm nf_1 \pm mf_2$, where n and m are integers. For the cases where $n \neq 0$ and $m \neq 0$, these components are called intermodulation (IM) distortion components.

The intermodulation components that appear first are of the 2nd and 3rd order. The 2nd order IM components at $f_1 \pm f_2$ typically lie outside of the desired signal band. It is important to note that the FM demodulation process itself constitutes a non-linear operation, and that the products $V_1(t)V_2(t-\tau)$ and $V_2(t)V_1(t-\tau)$ look like a second-order IM component. Second-order IM occurring in the LNA yields components around DC and at double the input frequency. These components are filtered out by the bandpass transfer of the LNA and don't do any harm.

However, the third-order IM components (IM3) at $2f_1 - f_2$ and $2f_1 - f_2$, lie in-band for a subcarrier FDMA system where users share the same RF center frequency. Third-order IM products from the LNA are negligible compared to the third order components generated by the active circuits of the FM demodulator, due to the larger signal amplitude at the demodulator input. The third-order intermodulation components are in-band and have a bandwidth 3 times as wide as the individual FM-UWB signals. In the next subsection, it will be shown that the system has already stopped working due to multiple-access interference before these IM components degrade performance.

Gain compression is the lowering of the small-signal gain by the presence of a large signal [15]. The small-signal gain is modulated by the larger signal and therefore its time average decreases. In a multi-user scenario, the detection of a weak signal may be compromised by a stronger interferer.

In the BAN application, interference from other FM-UWB users determines the large-signal requirements for the receiver front-end. It is important to define the worst-case interference scenario. In the FM-UWB BAN, this is a user only 50 cm away from the receiver. A maximum interference level of −58 dBm is obtained at an operating frequency of 7.5 GHz, 500 MHz bandwidth, antennas with 0 dBi gain and assuming free-space propagation.

The following subsections will further investigate the effect of the intermodulation components and gain compression.

5.2.4.1 Large-signal behavior of the FM demodulator

Since the wideband demodulator uses a bipolar Gilbert mixer as multiplier, it is useful to investigate the behavior of the hyperbolic tangent non-linearity y(x) defined by

$$y(x) = \tanh(x) = x - \frac{1}{3}x^3 + \frac{2}{15}x^5 - \frac{17}{315}x^7 + \dots \qquad (5.59)$$

Figure 5.18 shows the hyperbolic tangent function and its small-signal gain, A_V, defined by

$$A_V = \frac{dy}{dx} = 1 - \tanh^2(x), \qquad (5.60)$$

The small-signal gain is experienced by an infinitesimal sinusoidal input signal riding on top of the input signal x(t). Gain is defined at the fundamental frequency, as it would be measured by a frequency-selective signal analyzer.

Let the input signal be a single-tone sinusoidal signal with zero DC offset and frequency f_1, given by

$$x(t) = x_1(t) = A_1 \sin(2\pi f_1 t). \qquad (5.61)$$

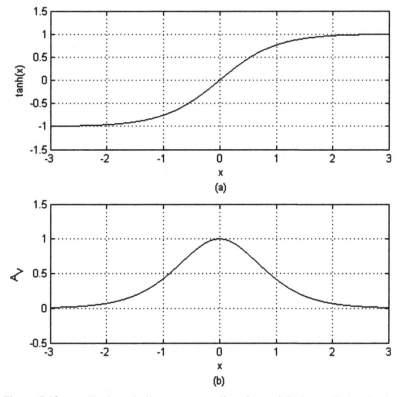

Figure 5.18 (a) The hyperbolic tangent non-linearity, and (b) its small-signal gain.

The small-signal gain is modulated by this signal. Its time average A_{Vavg}, for an overdrive O of A_1 is given by

$$A_{Vavg} = \int_0^{T_1} \left(1 - \tanh^2(O\sin(2\pi f_1 t))\right)dt, \qquad (5.62)$$

where $T_1 = 1/f_1$. This integral doesn't have an explicit solution, however, it can be evaluated numerically. For further calculations, an explicit expression $A_{v,avg}(O)$ is required. The integral is therefore approximated by the expression shown in Appendix D. This approximation yields results within 0.5 dB for overdrive values smaller than 10.

Figure 5.19 shows the input signal, x(t), small-signal gain, $A_V(t)$, and output signal, y(t), of the non-linear circuit for an overdrive equal to unity. The average small-signal gain has dropped to 0.67 (–3.5 dB). Figure 5.20 shows the small-signal gain versus overdrive. It should be noted that this is not the gain experienced by the (large) input signal x(t).

The gain for the larger signal at frequency f_2 is given by

$$A_{VL} = \frac{A_{OUT}}{A_{IN}} = \frac{1}{O}\frac{2}{T_2}\int_0^{T_2} \tanh(O\sin(2\pi f_2 t))\sin(2\pi f_2 t)dt, \quad (5.63)$$

with $T_2 = 1/f_2$.

Table 5.3 gives a number of useful values of overdrive and small-signal gain.

Next we consider the case of an input signal constituted by the sum of two sine waves: a small, wanted signal V_1 of amplitude A_1 and a large interfering signal V_2 of amplitude A_2, given by

$$V_i(t) = V_1(t) + V_2(t) = A_1\cos(2\pi f_1 t) + A_2\cos(2\pi f_2 t). \qquad (5.64)$$

The amplitudes are chosen to be $A_2 = 10A_1$. The overdrive O equals

$$O = A_1 + A_2 \approx A_2. \qquad (5.65)$$

Table 5.3 Small-signal gain versus overdrive for a hyperbolic tangent non-linearity

Overdrive	Overdrive, in dB	Small-signal Gain. in dB
0.33	–9.5	–0.5
0.50	–6	–1.0
1.0	0	–3.5
3.16	10	–13.3
10	20	–23.2

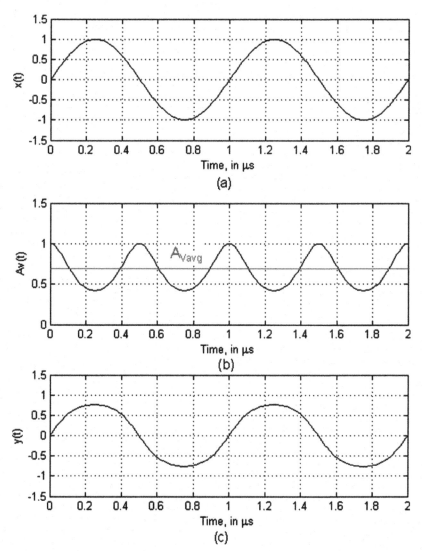

Figure 5.19 (a) Input signal, (b) small-signal gain, and (c) output signal for a hyperbolic tangent non-linearity with overdrive equal to unity.

Figure 5.21 shows the amplitude of the fundamental components (at frequencies f_1 and f_2) and the 3rd order intermodulation components IM3 (at frequencies $2f_1 - f_2$ and $2f_2 - f_1$) as a function of the overdrive O.

For a bipolar differential pair biased at tail current I_T and with collector resistors R_C, the output voltage V_O is given by

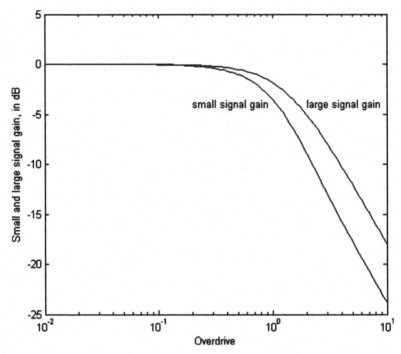

Figure 5.20   Small-signal and large-signal gain of a hyperbolic tangent non-linearity versus overdrive.

$$V_O = I_T R_C \tanh\left(\frac{V_i}{2V_T}\right). \qquad (5.66)$$

An overdrive of unity corresponds to a sine wave of amplitude $2V_T$ (50 ΩV), which represents a signal power of -16 dBm in a 50 W system. An overdrive of 0.5, or $A_2 = 25$ mV (-22 dBm in a 50 Ω system) yields 1 dB of gain compression. This level of compression may occur after the LNA in the FM-UWB receiver.

5.2.4.2 Large-signal behavior of the FM demodulator with additional preamplification

The multiplier gain K_D of the Gilbert multiplier with additional preamplification (see Figure 5.2) is given by (5.35). Gain compression occurs in both branches. The demodulator output voltage V_{DEM1} at subcarrier frequency f_{SUB1} is affected by the average small-signal gain A_{VNL1} and A_{VNL2} of the two non-linearities NL_1 and NL_2 respectively, which depend on the overdrives O_1 and O_2.

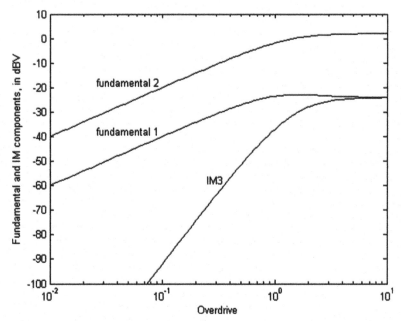

Figure 5.21 Amplitude of fundamental and IM3 components versus overdrive for a hyperbolic tangent non-linearity.

Overdrives O_1 and O_2 are equal to

$$O_1 = A_{VDEM1}O = \sqrt{\frac{A_{VDEMTOT}}{k_g}}O$$

and

$$O_2 = A_{VDEM2}O = \sqrt{k_g A_{VDEMTOT}}O. \qquad (5.67)$$

The demodulator output signal is proportional to

$$V_{DEM1} \propto A_{VDEM1}A_{VDEM2}A_{VNL1}(O_1)A_{VNL2}(O_2), \qquad (5.68)$$

where A_{VNL1} and A_{VNL2} model the gain compression occurring in each of the demodulator branches. For the bipolar Gilbert multiplier, (5.58) with $x = V_X/2V_T$ or $x = V_Y/2V_T$, describes the non-linearity of each differential pair.

For a given demodulator gain, $A_{VDEMTOT}$, we dare to determine the best gain distribution between A_{VDEM1} and A_{VDEM2} for minimum gain compression.

For various values of the additional demodulator gain, Figure 5.22 shows the overdrive (in dB) for which the demodulator output voltage shows 1 dB of compression. The data shows that a k_g of one, i.e., $A_{VDEM1} = A_{VDEM2}$ yields the lowest value, i.e., best large-signal performance. It can be seen that an increase of $\Delta A_{VDEMTOT}$ dB of the total demodulator gain, and $\Delta A_{VDEMTOT}/2$ dB in both A_{VDEM1} and A_{VDEM2} lowers the overdrive value by $\Delta A_{VDEMTOT}/2$ dB for -1 dB compression.

A Practical Example

Adding gain in a single branch of the demodulator is not advantageous for the gain compression characteristics. As mentioned in Section 5.1.3.1, the FM demodulator in [10] has 26 dB of gain in the delayed path (i.e., $A_{VDEM2} = 20$ and $k = 20$). The demodulator sensitivity is increased by 13 dB because of the extra gain, however, the 1 dB compression point is lowered by 23 dB, thereby

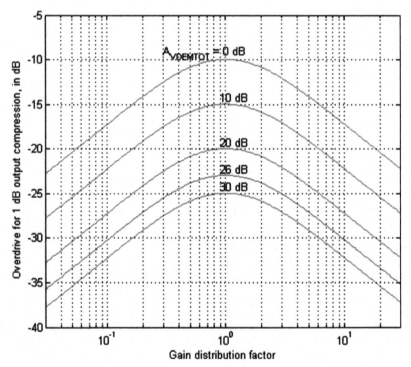

Figure 5.22   Overdrive corresponding to 1 dB of demodulator output signal compression versus gain distribution factor.

lowering the dynamic range by 10 dB. Knowing that an overdrive of unity for a bipolar differential pair corresponds to –16 dBm, the –1 dB compression point at the input of the FM demodulator is now at –49 dBm. Inspection of curve 3 in Figure 5.17 shows that a LNA gain of 13 dB is required to obtain a receiver sensitivity of –85 dBm. The –1 dB compression point at the receiver input therefore becomes –62 dBm.

To determine if this value is sufficient, we assume the worst-case interference scenario constituted by a user at 50 cm distance from the receiver, yielding an interference level of –58 dBm. This is 4 dB above the –1 dB compression point, resulting in a few dB of additional compression in the wanted signal.

However, with an interferer at –58 dBm, the minimum detectable signal level at the receiver input is also determined by the multiple-access interference as shown in Section 2.5.1. Processing gain in a 250 kbps system with 500 MHz bandwidth is 30 dB (see Figure 2.21), meaning that a 20 dB stronger interferer will lower the probability of error to 1×10^{-3}. When the interferer is at –58 dBm, the weak signal needs to be stronger than –78 dBm to be detected. When such a strong interferer is detected, the LNA gain can be lowered by 5 dB. The input sensitivity changes to –81 dBm and the –1 dB compression point is then at –57 dBm, which is just good enough.

The way to improve the large-signal behavior of the receiver front-end described in [5] is to create unity gain ($A_{VDEM2} = 1$) in the delayed branch of the wideband demodulator and increase the LNA gain. The diverted current from the multiplier should then be used to bias the delay stages for unity gain, and use the remaining current in an intermediate preamplification stage.

5.2.4.3 Intermodulation components in the FM demodulator

We will now examine the case of two FM-UWB signals at the receiver input (V_1 and V_2) and determine whether the third-order intermodulation components created in the FM demodulator, which are in-band, are harmful.

As already stated in Section 3.2.1 of this book, the multiple access (MA) residue originating from the terms $V_1(t)V_2(t-\tau)$ and $V_2(t)V_1(t-\tau)$ limits the radio performance in a multi-user environment where subcarrier FDMA is employed.

The third-order intermodulation components generated by the differential pairs of the multiplier mix with both V_1 and V_2. They yield a wideband signal around DC. Part of the signal spectrum lies within the subcarrier bandwidth.

In order to determine whether the IM3 components are harmful, their power needs to be compared with the power of the multiple access components. As shown in Figure 5.21, the IM3 components are smaller than the output signal due to the smaller input signal, V_1. Therefore the multiple-access interference due to $V_1 \times V_2$ will always be larger than $IM3 \times V_2$, and the IM3 components may be ignored.

5.3 Low Noise Amplifier Implementation

Various circuit techniques may be exploited to boost the LNA performance and lower its power consumption. Figure 5.23 shows a simplified schematic of the two-stage LNA published in [10]. The two-stage LNA has a single-ended input and is designed for a 50 Ω antenna and prefilter. It is intended for

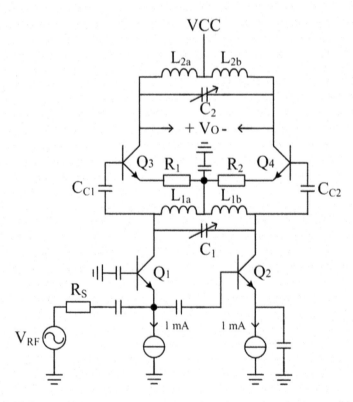

Figure 5.23   Simplified schematic of LNA exploiting noise cancelling and current reuse.

operation at 7.45 GHz and has 500 MHz bandwidth. Detailed schematics can be found in Appendix C.

The single-ended-to-differential conversion is implemented by common-base (CB) and common-emitter (CE) stages with paralleled inputs [16]. The inverting common-emitter (CE, Q_2) and non-inverting common-base (CB, Q_1) stages generate an accurate differential output signal. The collector shot noise from transistor Q_1 is cancelled in the differential output current when $g_{m2}R_S = 1$, where g_{m2} is the transconductance of transistor Q_2 and R_S is the source impedance [17]. Since common base transistor Q_1 is biased at 1 mA, the LNA input impedance equals $1/g_{m1} = 25$ Ω. An additional matching network consisting of a bondwire, the package lead and a PCB trace transform this value into 50 Ω. The output current of the first stage is transformed into a voltage by the parallel resonant load $L_1 - C_1$. The first stage is followed by a differential CE stage constituted by transistors Q_3 and Q_4 with resonant load $L_2 - C_2$. The overall voltage gain is 30 dB. The bias current of the input stage is reused by the second stage to lower power consumption. The RF signal is coupled to the second stage by capacitors C_{C1} and C_{C2}. Capacitors C_1 and C_2 are differential varactor diodes that can be tuned independently. This allows for the use of staggered tuning to increase the LNA bandwidth [18].

5.4 Receiver Subcarrier Processor Implementation

Figure 5.24 shows a representative example of the signal present at the input of the subcarrier processing block, which are two FSK subcarriers with frequencies of 1 MHz and 1.5 MHz. The noise floor is the result of multiple-access interference. Due to frequency-selective multipath, harmonics of the subcarrier signals are also present. Measures have to be taken to insure that harmonics from one subcarrier can't degenerate the performance of another one.

The subcarrier processor design is constrained by selection of the wanted FSK subcarrier, the noise bandwidth, and attenuation of signals at subcarrier harmonics.

Figure 5.25 shows the block diagram of the subcarrier processor. Direct-conversion to baseband is used to relax the subcarrier channel filter specifications. After analog lowpass filtering, the subcarrier signal is amplified, hardlimited (1-bit analog-to-digital conversion) and demodulated in a digital FSK demodulator. It is possible to choose the boundary between the analog and digital signal processing domains closer to the antenna [19].

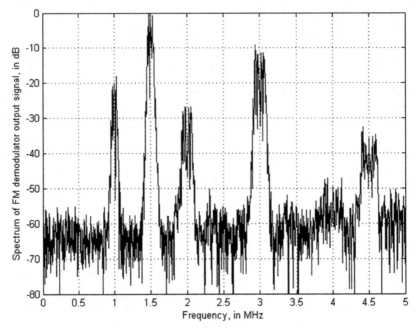

Figure 5.24   Typical input signal for the subcarrier processing.

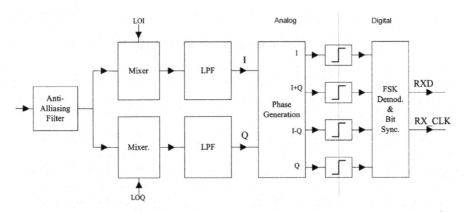

Figure 5.25   Block diagram of receiver subcarrier processor.

Triangular LO signals (LOI and LOQ in Figure 5.24) drive the switching quads of the two Gilbert downconversion mixers for additional attenuation of odd harmonics [20]. The hyperbolic tangent transfer function of a differential pair transforms the triangle input signal into a good approximation of a sine wave [21]. This yields 30 dB of attenuation for signals at the 3rd and 5th harmonic of the LO frequency. The triangular, quadrature LO signals are generated by direct digital synthesis (DDS) as used in the transmitter. As seen in Chapter 4, the use of a triangular subcarrier waveform has advantages for the transmitter. The necessity of triangular LO signals in the receiver is the decisive factor in favor of a triangular DDS signal waveform.

It was found by practical experiments that harmonics may be as strong as the fundamental component. Since a 20 dB stronger interferer may occur, yielding a 40 dB stronger subcarrier signal, the 30 dB of attenuation provided by the mixers is insufficient. A minimum of 23 dB of additional attenuation needs to be provided by the anti-aliasing filter (AAF) for frequencies above 3 MHz to keep the signal-to-interference ratio above 13 dB. The attenuation of the AAF must be low between 1 and 2 MHz and high for frequencies greater than three times the lowest subcarrier frequency.

The signals at the subcarrier processor input are in the 10 μV to 10 mV range, which corresponds to 60 dB dynamic range. Table 5.4 shows the gain distribution inside the subcarrier processor. Switchable gain in the AAF lowers the dynamic range required for the subsequent blocks. When the input signal level is above 100 μV, the AAF gain is reduced by 20 dB. The dynamic range of the AAF output signals is reduced by 20 dB and the signal values are between 40 μV and 4 mV. After the mixers, the signal values are between 160 μV and 16 mV.

5.4.1 Anti-aliasing Filter Implementation Example

The AAF is a fourth-order elliptic lowpass filter with 2 MHz cut-off frequency and a zero in its transfer function at 3.5 MHz, which attenuates the harmonics of the subcarrier signals. Figure 5.26 shows its transfer function. The filter has two gain settings: −8 and +12 dB [22, 23]

Table 5.4 Gain distribution of subcarrier processor

Block	Gain, in dB
AAF	−8/12*
Mixer	12
LPF	0
Phase generation	0
Limiter amplifiers	80
Total	86/104*

*switchable gain in the AAF.

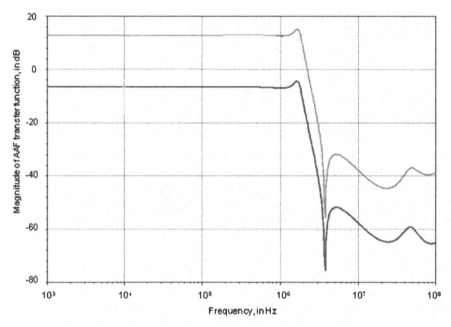

Figure 5.26   Simulated AAF filter transfer function [23].

5.4.2 Lowpass Filter Implementation Example

The lowpass filters after the mixers have a cut-off frequency f_{LP} equal to half the subcarrier bandwidth, given by

$$f_{LP} = \frac{B_{SUB}}{2} = \frac{R}{2}(\beta_{SUB} + 1). \tag{5.69}$$

Figure 5.27 illustrates the subcarrier selectivity obtained with a fifth-order 1 dB ripple Chebyshev lowpass filter with a cut-off frequency of 100 kHz. This filter provides 60 dB of attenuation for the adjacent subcarrier frequency. Assuming a signal-to-interference ratio (SIR) of 20 dB for correct demodulation, this yields 40 dB of margin at the subcarrier level and 20 dB at RF because of the quadratic transfer function of the wideband FM demodulator. For a short-range BAN application, 20 dB of margin yields adequate performance most of the time. Steeper filtering doesn't make sense because the system is multiple-access limited when various users employ the same RF carrier frequency and exploit subcarrier FDMA.

The lowpass filter was implemented as a differential gm-C filter to reduce chip area.

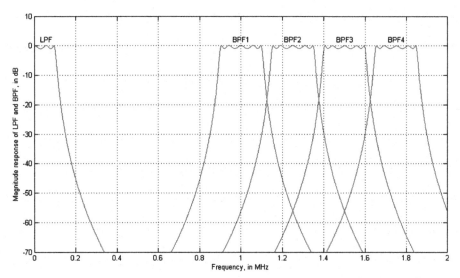

Figure 5.27 Equivalent subcarrier filter selectivity (BPF1 – BPF4) obtained with a 5th order 1 dB ripple Chebyshev lowpass filter (LPF).

5.4.3 FSK Demodulator and Bit Synchronizer

After the lowpass filters, the baseband I and Q signals are combined to yield 4 phases (I, Q, I + Q and I – Q). The hardlimited versions of these four signals are fed into a digital FSK demodulator that determines the sign of the instantaneous frequency at each zero crossing of the four input signals [24].

The raw demodulated data (RXD_RAW) is lowpass filtered in a 16-tap FIR filter and fed into a bit synchronizer circuit that delivers the bit clock (RX_CLK) and retimed data (RXD) at its output. The bit synchronizer circuit is a digital delay-locked loop (DLL) with an adaptive loop bandwidth operating at 16 times the bit rate. Figure 5.28 shows the block diagram of the bit synchronizer. The clock oscillator provides a clock at the bit rate whose phase needs to be adjusted to the input raw data (RXD_RAW). A phase detector measures the error of the clock phase at each data transition. Its output signal is lowpass filtered and used to adjust the phase of the clock.

The phase shifter operates at 16 times the bit rate, and the output clock RX_CLK can have 16 discrete phase values (0: π/8: 15π/8). At each transition, the clock phase is adjusted until it enters a limit cycle around its equilibrium position.

A loop filter with adaptive bandwidth allows for rapid acquisition and good tracking performance [25, 26]. Details on the measured performance of the receiver can be found in Section 6.3.

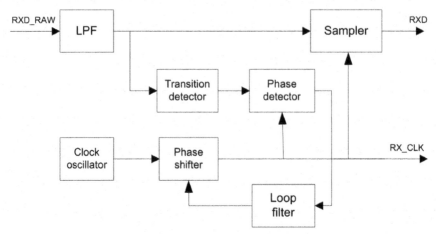

Figure 5.28　Bit synchronizer block diagram.

5.5 Conclusions

This chapter has investigated practical implementation of the FM-UWB receiver. The performance of the RF front-end is strongly influenced by the wideband FM demodulator. The difficulties are in the implementation of a flexible (i.e., tunable) time delay circuit and a high-gain, low-noise multiplier circuit.

The tunable delay circuit is advantageously implemented by a cascade of a lattice bandpass filter that yields –90 degrees of phase shift at its resonant frequency, and a simple bandpass filter. The double-balanced Gilbert multiplier makes a good choice for the multiplier circuit and its noise can be lowered by current bleeding, which yields a sensitivity improvement. The diverted current can then be reused to bias gain stages in the delay circuit and the LNA.

Adding preamplification to the delay circuit of the FM demodulator lowers the gain compression point more than the sensitivity is increased, yielding reduced dynamic range. The gain in the delay circuit should therefore be chosen equal to unity. Adding gain in the LNA, or adding an intermediate amplifier between the LNA and FM demodulator improves the sensitivity with a minimum penalty on the large-signal behavior. Automatic gain control (AGC) may be applied to further increase the dynamic range.

Various circuit techniques may be exploited to improve the LNA performance and lower its power consumption. A two-stage amplifier exploiting noise cancelling, staggered tuning and current reuse has been presented.

The subcarrier processor uses a direct-conversion approach. However, due to the presence of harmonics of the subcarriers at the FM demodulator output, additional filtering is required prior to downconversion to baseband of the FSK subcarrier signal. This is performed by an analog lowpass filter. Additional attenuation of the odd harmonics by 30 dB is obtained by using a triangular LO signal in the downconversion mixers.

References

[1] IEEE 802.15 WPAN^{TM} Task Group 6 Body Area Networks (BAN), website URL: http://www.ieee802.org/15/pub/TG6.html.

[2] K. Y. Yazdandoost, K. Sayrafian-Pour, *Channel Model for Body Area Network (BAN)*, IEEE 802.15-08-0780-09-0006, 27 April 2009. [Online]. Available: https://mentor.ieee.org/ 802.15/dcn/08/15-08-0780-09-0006-tg6-channel-model.pdf.

[3] John Farserotu, John Gerrits, Jérôme Rousselot, Gerrit van Veenendaal, Manuel Lobeira, John Long, *CSEM FM-UWB Proposal*, IEEE P802.15 Working Group for Wireless Personal Area Networks (WPANs), Task Group TG6 Body Area Networks (BAN), May 2009, Montreal. Doc.:IEEE802.15-09-0276-00-0006, [Online]. Available: https://mentor.ieee.org/802.15/dcn/09/15-09-0276-00-0006-csem-fm-uwb-proposal.pdf.

[4] Krijn Leentvaar and Jan H. Flint, "The Capture Effect in FM Receivers," *IEEE Transactions on Communications*, vol. 25, no. 4, pp. 531–539, May 1976.

[5] Kuo-Ken Huang, Meng-Ju Chiang, and Ching-Kuang. C. Tzuang, "A 3.3mW-band 0.18 m1p6m CMOS active bandpass filter using complementary current-reuse pair," IEEE Microwrowave Wireless Compononent Letters, vol. 18, no. 2, pp. 94–96, February 2008.

[6] Sen Wang, Kun-Hung Tsai, Kuo-Ken Huang, Si-Xian Li, Hsien-Shun Wu, and Ching-Kuang C. Tzuang, "Design of X-Band RF CMOS Transceiver for FMCW Monopulse Radar", IEEE Transactions on Microwave Theory and Tecniques, vol. 57, no. 1, pp. 61–70, January 2009.

[7] Gunhee Han and Edgar Sanchez-Sinencio, "CMOS Transconductance Multipliers: A Tutorial," *IEEE Transactions on Circuits and Systems II: Analog and Digital Signal Processing*, vol. 45, no. 12, pp. 1150–1563, December 1998.

[8] John F.M. Gerrits, John R. Farserotu, John R. Long, "A Wideband FM Demodulator for a Low-Complexity FM-UWB Receiver," *Proc. ECWT 2006*, pp. 99–102.

[9] Anatol I. Zverev, *Handbook of Filter Synthesis*, New York: John Wiley & Sons, 1967.

[10] Y. Zhao, Y. Dong, J.F.M. Gerrits, G. van Veenendaal, J.R. Long and J.R. Farserotu,"A Short Range, Low Data Rate, 7.2G–7.7GHz FM-UWB Receiver Front-End," *IEEE Journal of Solid-State Circuits*, vol. 44, no. 7, pp. 1872–1882, July 2009.

[11] Barrie Gilbert, "A precision four-quadrant multiplier with subnano-second response," *IEEE Journal of Solid-State Circuits*, vol. 3, No. 4, pp. 365–373, December 1968.

[12] Alberto Bilotti, "Applications of a Monolithic Analog Multiplier," *IEEE Journal of Solid-State Circuits*, vol. 3, no. 4, pp. 373–380, Dec. 1968.

[13] Paul R. Gray, Paul J. Hurst, Stephen H. Lewis, Robert G. Meyer, *Analysis and Design of Analog Integrated Circuits*, Fourth Edition, New York, John Wiley and Sons, 2001.

[14] Kevin McClaning, Tom Vito, *Radio Receiver Design*, Atlanta: Noble Publishing, 2000.

[15] Robert G. Meyer and Alvin K. Wong, "Blocking and desensitization in RF amplifiers," *IEEE Journal of Solid-State Circuits*, vol. 30, no. 8, August 1995, pp. 944–946.

[16] Bram Nauta, *Single-to-differential converter*, US patent 5,404,050, April 4, 1995.

[17] Federicio Bruccoleri, Eric A.M. Klumperink, and Bram Nauta, "Wide-Band CMOS Low-Noise Amplifier Exploiting Thermal Noise Cancel-ing," *IEEE Journal of Solid-State Circuits*, vol. 39, no. 2, pp. 275–282, February. 2004.

[18] M.M. Mcwhorter, J.M. Pettit, "The Design of Stagger-Tuned Double-Tuned Amplifiers for Arbitrarily Large Bandwidth," *Proc. of the IRE*, vol. 43, no. 8, pp. 923–931, August 1955.

[19] Robert Bogdan Staszewski, Khurram Muhammad, Dirk Leipold, Chih-Ming Hung, Yo-Chuol Ho, John L. Wallberg, Chan Fernando, Ken Maggio, Roman Staszewski, Tom Jung, Jinseok Koh, Soji John, Irene Yuanying Deng, Vivek Sarda, Oscar Moreira-Tamayo, Valerian Mayega, Ran Katz, Ofer Friedman, Oren Eytan Eliezer, Elida de-Obaldia, and Poras T. Balsara, "All-Digital TX Frequency Synthesizer and Discrete-Time Receiver for Bluetooth Radio in 130-nm CMOS,"

IEEE Journal of Solid-State Circuits, vol. 39, no. 12, pp. 2278–2291, December 2004.

[20] Dieter Janta, Wolfgang Nolde, *Demodulation circuit having triangular wave for blocking harmonics*, US patent 4,560,942, December 24, 1985.

[21] Robert G. Meyer, Willy M.C. Sansen, Sik Lui, and Stefan Peeters, "The Differential Pair as a Triangle-Sine Wave Converter," *IEEE Journal of Solid-State Circuits*, vol. 11, no. 6, pp. 418–420, June 1976.

[22] John Gerrits, Marco Giardina, Manuel Lobeira, Jan Mikkelsen, Peter Nilsson, Dominique Noguet, Tian Tong. Magnet Beyond deliverable D5.2.2 & D5.2.3, *IC Design Report*, October 2006.

[23] Hamid Bonakdar, John Gerrits, Dominique Noguet, Gerrit van Veenendaal, Yi Zhao, Magnet Beyond deliverable D5.4.2b, *LDR High Band Test Report*, September 2008.

[24] Ian A.W. Vance, *Decoding logic for frequency shift keying receiver*, US patent 4,322,851, March 30, 1982.

[25] A. Erbil. Payzin, "Analysis of a digital bit synchronizer," *IEEE Transactions on Communications*, vol. 31, no. 4, pp. 554–560, April 1983.

[26] Helmuth Brügel, and Peter F. Driessen, "Variable Bandwidth DPLL Bit Synchronizer with Rapid Acquisition Implemented as a Finite State Machine," *IEEE Transactions on Communications*, vol. 42, no. 9, pp. 2751–2759, September 1994.

6

Measured Performance of FM-UWB

This chapter presents measurement results made on a hardware prototype operating at a center frequency of 7.45 GHz that was developed in the Sixth Framework IST project 507102, MAGNET Beyond [1]. Dedicated ICs have been manufactured for both transmitter and receiver. Figure 6.1 shows from the left to the right: RF transmitter, receiver front-end and subcarrier processor ICs.

The next subsections present the transceiver prototype and its measured performance. The measured performance is compared to the theory presented in earlier chapters. Section 6.1 presents the FM-UWB transceiver prototype and its characteristics. Section 6.2 shows measurement results from the transmitter and Section 6.3 presents measurement results of the receiver building blocks. Section 6.4 presents measurement results made on the complete transceiver. Section 6.5 closes the chapter with conclusions.

6.1 Transceiver Prototype

Figure 6.2 shows the transceiver prototype [2]. It consists of two boards. The first board contains the analog parts of the FM-UWB radio: the RF transmitter, the receiver front-end and the subcarrier processor ICs shown in Figure 6.1, plus additional off-the-shelf components, such as the DDS DACs and a few linear regulators to generate the various supply voltages. The second board band contains an Altera MAX II CPLD that implements the DDS, the FSK demodulator and bit synchronizer, control logic and the communications protocol. The FM-UWB radio has various operating modes. It can be configured for BER measurements using a pseudo random test pattern. Single errors can be inserted into the sequence to obtain a known bit error rate value. In a second operating mode, the transmitter continuously sends a string of ASCII characters. In a third operating mode, the radio acts as wireless modem between an external data source, e.g., a pulse oximetry (SpO_2) sensor with RS-232 output [3] and a PC for data visualization.

133

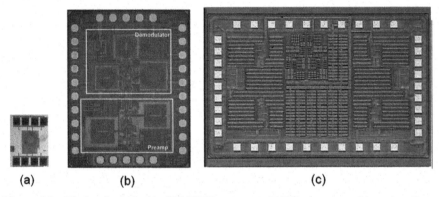

(a)   **(b)**   **(c)**

Figure 6.1   ICs developed for the FM-UWB prototype: (a) RF transmitter, (b) receiver front-end, (c) subcarrier processor.

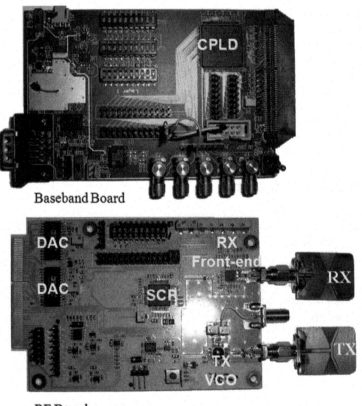

Figure 6.2   FM-UWB transceiver prototype.

Table 6.1   FM-UWB prototype characteristics [2]

Parameter	Value
RF center frequency	6.5–7.6 GHz
RF bandwidth	500 MHz
RF output power	(−15)–(−9) dBm
Subcarrier frequency	1–2 MHz
Subcarrier modulation	2-FSK
Raw data rate	31.25, 62.5, 125, 250 and 50, 100, 200 kbps

Separate transmit and receive antennas are employed for maximum flexibility. The antenna used in the measurements is a UWB bowtie antenna covering the 3–10 GHz range with a notch at 5.2 GHz [4]. The measured antenna gain is 1.5 dBi at 7.5 GHz. Table 6.1 presents an overview of the FM-UWB prototype characteristics. For historical reasons, two different sequences of bit rate values have been implemented.

6.2 Transmitter Performance

The transmitter comprises the DDS-based subcarrier signal generation and the RF signal generation as described in Chapter 4. The DDS code runs on an Altera MAX II CPLD from a 32 MHz clock and employs a 16-bit wide phase accumulator driving two off-the-shelf 10-bit DACs. The RF section of the transmitter consists of an LC-VCO followed by a cascaded inverters output stage driving the antenna [5]. It has been implemented in UMC 1P6M 0.13 μm RF-CMOS technology. The next subsections present measurement results of the subcarrier signal and the FM-UWB signal.

6.2.1 Subcarrier Signal

Figure 6.3 shows the measured spectrum of the subcarrier signal at 1 MHz, subcarrier deviation is 50 kHz and the data rate is 31.25 kbps. A 511 bit long pseudo-random data sequence is used. There is room to improve rejection (44 dB) of the first sidelobes appearing at 250 kHz from the FSK center frequency.

6.2.2 FM-UWB Signal

The measured VCO tuning curves for supply voltages from 1.0 V to 1.2 V are shown in Figure 6.4. Output power versus frequency for various VCO and output buffer supply voltage configurations are shown in Figure 6.5.

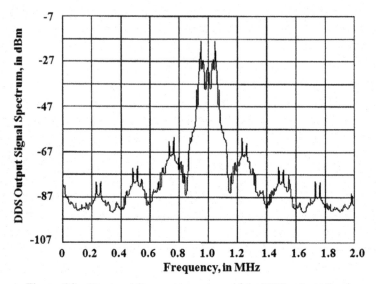

Figure 6.3   Measured frequency spectrum of the DDS output signal.

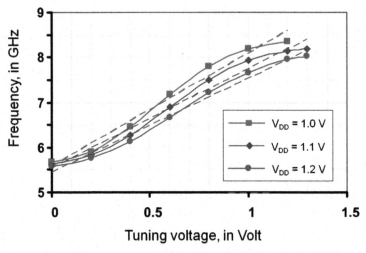

Figure 6.4   Measured VCO tuning curves [5].

In Section 4.2, it was stated that for achieving not more than 1 dB of peaking of the PSD of the FM-UWB signal, the VCO gain at the band edges should not be lower than 80% of the value at the VCO center frequency. For a 500 MHz signal bandwidth, this target is achieved for a center frequency range between 6.5 and 7.6 GHz.

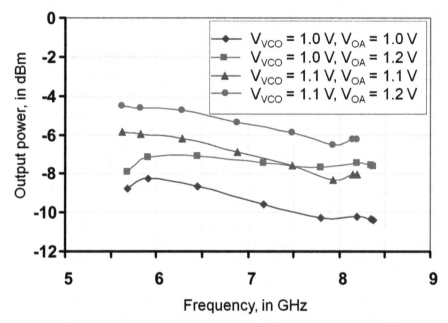

Figure 6.5 Measured transmitter output power versus frequency [5].

Phase noise performance of approximately –107 dBc/Hz at 1 MHz offset is obtained at the highest operating frequency, with overall power consumption of 2.5 mW at 1 V supply as illustrated in Figure 6.6. Note that the measured value occurs at the limit of the region where the phase noise measurement data are trustworthy. Actual phase noise could be up to 5 dB higher [6]. Figure 6.7 shows the spectrum of the FM-UWB signal for a 1 MHz subcarrier frequency.

The prototype was also used for pre-compliance testing at the facilities of EMC Montena in Rossens, Switzerland. Figure 6.8 shows the measured spectrum result which is FCC compliant. A possible cause for the ripple observed on the spectrum analyzer trace could be impedance mismatch between the receive antenna and spectrum analyzer.

6.3 Receiver Performance

The receiver comprises the RF front-end, (LNA with FM demodulator) and subcarrier processor. The next subsections present measurement results made on the front-end and the subcarrier processor.

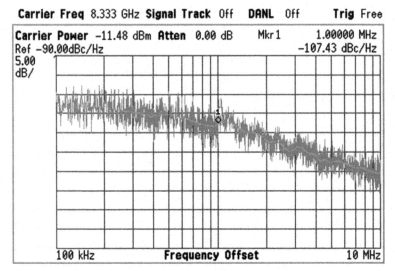

Figure 6.6   Measured VCO phase noise at 8.33 GHz [5].

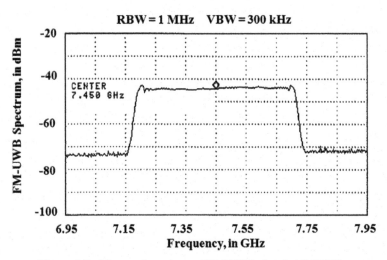

Figure 6.7   Measured spectrum of FM-UWB signal at 7.45 GHz.

6.3.1 Front-end Measurements

The FM-UWB receiver front-end is described in [7]. It comprises the low noise amplifier and FM demodulator. Target power consumption of the receiver front-end demonstrator was 10 mW from a 1.8 V supply. The prototype

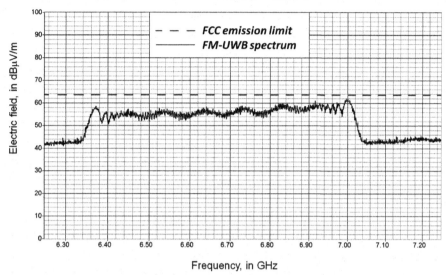

Figure 6.8 Measured FM-UWB signal and FCC emission limit.

was fabricated in a 0.25 μm SiGe:C-BiCMOS technology (NXP Semi-conductors' QUBiC4X) [8]. Detailed schematics of the LNA and FM demod-ulator can be found in Appendix C. Measurements were made on a stand-alone version of the LNA with emitter follower output amplifiers yielding 100 Ω differential output impedance. Figure 6.9 shows the S_{21} values measured with a balun at the LNA output. The LNA voltage gain is 9 dB higher than the measured S_{21} value, 6 dB originates from the output buffers and 3 dB from the 50–100 Ω balun. More than 30 dB of voltage gain is obtained at 7.45 GHz, with a –3 dB bandwidth of 600 MHz. The figure also illustrates the staggered tuning capabilities of this two-stage LNA. Figure 6.10 shows the measured LNA noise figure which is lower than 6 dB from 7.2 to 8 GHz.

The combined transfer of the LNA and FM demodulator has also been measured. The FM-UWB transmitter was used to generate the FM-UWB signal. A 500 kHz triangular signal is modulating the transmitter VCO. Figure 6.11 shows a photo of this wired setup. A resistive splitter and a variable attenuator are connected between the transmitter and receiver. A low-frequency buffer amplifier is inserted between the FM demodulator output and the oscilloscope. Figure 6.12 shows the modulating signal and

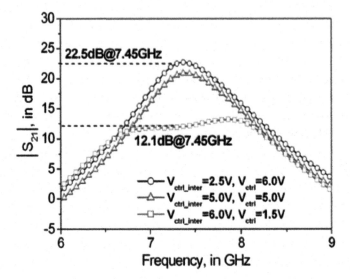

Figure 6.9   Measured S_{21} of the LNA [7].

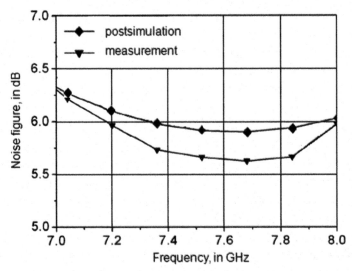

Figure 6.10   Measured LNA noise figure [7].

the FM demodulator output signal for an FM-UWB signal with 500 MHz
and 1 GHz bandwidths. RF input level is –56 dBm. The demodulator was
tuned for 1 GHz bandwidth as in the example presented in Section 5.2.1.3.

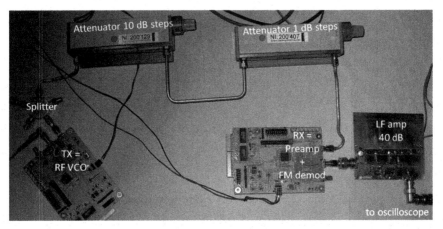

Figure 6.11 Setup for measuring the FM demodulator transfer.

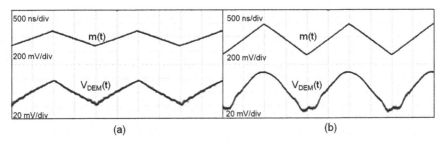

Figure 6.12 Measured modulating signal (m(t)) and FM demodulator output signal ($V_{DEM}(t)$) for an RF bandwidth equal to (a) 500 MHz and (b) 1 GHz.

6.3.2 Subcarrier Processor Measurements

Measurements were performed to test the subcarrier processor IC [9]. Figure 6.13 shows the measured transfer function of the anti-aliasing filter (AAF). The filter has a gain of 12 dB. Its –3 dB bandwidth is 2.05 MHz and the attenuation at 3 MHz equals 26 dB, which is consistent with the specifications given in Section 5.4.

Figure 6.14 shows the measured transfer function of the lowpass filter adjusted for a cut-off frequency of 100 kHz. A comparison with the filter characteristic shown in Figure 5.27, shows good agreement. The attenuation at two times the cut-off frequency equals 46 dB in both cases. At higher frequencies, the filter transfer function becomes flat, which will lower the attenuation of the adjacent subcarrier signal and impact the multi-user performance. In Section 5.2.2, it was shown that for short-range BAN applications, 60 dB of

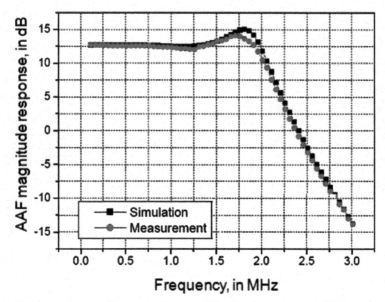

Figure 6.13   Simulated and measured anti-aliasing filter transfer function.

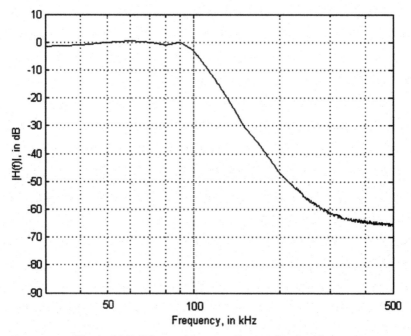

Figure 6.14   Measured lowpass filter transfer function.

attenuation at the adjacent subcarrier frequency yields adequate performance most of the time. There is room to improve the performance of this subcarrier channel filter.

6.4 Overall Transceiver Measurements

This subsection presents measurements made on the complete FM-UWB transceiver: bit error rate (BER) measurements with AWGN, link span, multi-user performance, performance with narrowband interference and receiver synchronization time.

6.4.1 BER Measurements with AWGN

The results of wired BER measurements made on 4 different receiver prototypes are shown in Figure 6.15. The black solid line is the analytical reference curve for FM-UWB modulation. The colored dashed lines represent the measurement results. Measurement time for the lower BER values was 4 minutes, which corresponds to 15 million transmitted bits. The measured FM-UWB receiver sensitivity for a data rate of 62.5 kbps is –87 dBm for

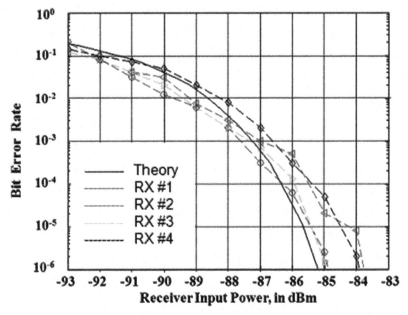

Figure 6.15 BER measurement results made on 4 receivers at 62.5 kbps.

a BER of 1×10^{-3}, whereas at -85 dBm is required to reach a BER of 1×10^{-6}. The four receivers show very similar performance. As described in Section 3.1, the receiver was designed for a sensitivity of -85 dBm at a data rate of 250 kbps and a BER of 1×10^{-6}. Using (3.13), it can be shown that sensitivity at 62.5 kbps should be 3.7 dB better, i.e., -88.7 dBm. The difference may be partially explained by the fact that the FM demodulator was tuned for a bandwidth of 1 GHz, whereas the FM-UWB signal has a bandwidth of 500 MHz, which corresponds to an overdrive of 0.5. As shown in Section 2.3.1, operation at overdrives lower than one yields a sensitivity loss.

6.4.2 Link Span

Propagation experiments were carried out in a 2×5 m^2 laboratory environment and a 20×10 m^2 commercial cafeteria. Propagation in these environments is predominantly line of sight (LOS). The laboratory has metallic walls and objects that create multipath fading, whereas the cafeteria is a more open space. The link margin was measured at various distances by inserting additional attenuation between antenna and receiver until the BER degraded to 1×10^{-3}. Figure 6.16 presents the results. The range that can be covered in the cafeteria with a transmit power of -14.3 dBm is approximately 10 meters. The better propagation in the laboratory environment can be explained by its metallic walls that act like a waveguide. Propagation is better than free-space attenuation in this environment. The link margin calculated for a 250 kbps link with a BER equal to 1×10^{-6}, as presented in Section 5.1, is 7.2 dB. For a BER equal to 1×10^{-3}, a data rate of 62.5 kbps, and no TX backoff, this number increases to approximately 15 dB at 3 meter distance.

6.4.3 Multi-user Performance

Tests were also carried out to validate the concept of subcarrier FDMA, where different users share the same RF bandwidth and distinguish themselves by their subcarrier frequencies. The system was operated at 50 kbps with 50 kHz subcarrier deviation and a subcarrier spacing equal to 250 kHz. Figure 6.17 shows the signal after the wideband FM demodulator with the FM-UWB interferer power 10 dB stronger than the desired signal at the receiver input (i.e., -60 dBm subcarrier at 1.5 MHz versus -70 dBm subcarrier at 1.25 MHz). The amplitude of the interfering subcarrier is 20 dB stronger than the desired one, which is due to the quadratic characteristic of the FM demodulator. Also, the noise floor increases due to multiple-access

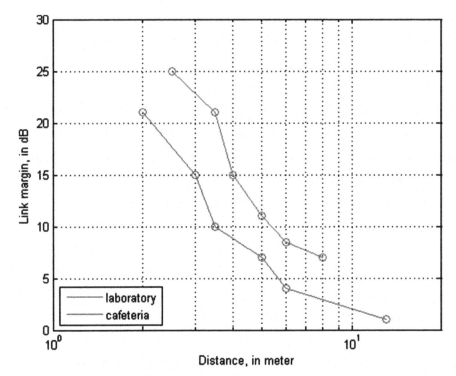

Figure 6.16 Measured link margin at 62.5 kbps in laboratory and commercial cafeteria environments.

interference. The measured BER drops to 1×10^{-3} for a signal-to-interference ratio (SIR) of –14 dB (i.e., interferer power is 14 dB stronger than the desired signal).

As stated in Section 3.2.4, when using a subcarrier FDMA scheme, there are four reasons why a wanted signal may be drowned by a stronger (interferer) signal using the same RF center frequency and a different subcarrier frequency: multiple-access interference, insufficient sidelobe rejection in the transmitter subcarrier generation, transmitter RF oscillator phase noise, and insufficient selectivity of the receiver subcarrier filtering. The processing gain for this 50 kbps system is 37 dB which would put the fundamental limitation at a 27 dB stronger interferer. This interferer subcarrier level prior to FSK demodulation should be at least 10 dB lower than the wanted subcarrier level.

In this prototype, the performance appears to be limited by the sidelobe rejection (44 dB, see Figure 6.3) of the subcarrier generation. The first sidelobe

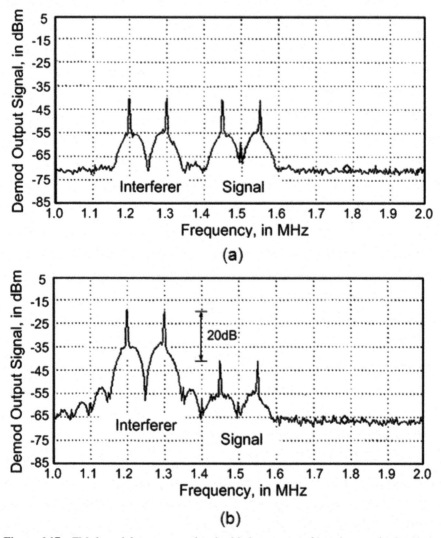

Figure 6.17 FM demodulator output signal with 2 users (a) of equal strength, (b) 10 dB difference.

of the interferer subcarrier signal shows up as co-channel interference in the adjacent subcarrier channel. Improving the rejection of the first sidelobe by 10 dB will yield 5 dB more margin at RF. The next limiting factor is the subcarrier channel selection filter, which attenuates the adjacent subcarrier by 55 dB at 250 kHz (see Figure 6.14).

6.4.4 Narrowband Interference

The 50 kbps FM-UWB system was also measured with narrowband interference. The FM-UWB signal power was −80 dBm and the CW interferer was 20 dB stronger (−60 dBm). Figure 6.18 shows the two signals as measured on the spectrum analyzer. In order to see the FM-UWB signal, the spectrum analyzer was connected to a 20 dB stronger version of the signal present at the FM-UWB receiver input. A signal power of −40 dBm on the spectrum analyzer corresponds to a signal of −60 dBm at the receiver input. It was found that the BER degraded to 1×10^{-3} with the 20 dB stronger narrowband interferer placed in the middle of the FM-UWB spectrum. According to the theory presented in Section 3.2.1, this should occur for an interferer that is 24 dB stronger than the wanted signal.

6.4.5 Receiver Synchronization Time

Figure 6.19 illustrates the receiver synchronization time for a data rate of 31.25 kbps (i.e., a symbol time of 32 µs). At the rising edge of the TX_ENABLE signal, the transmitter starts transmitting data. The raw data (RXD_RAW) is available instantaneously and a well-aligned receiver data

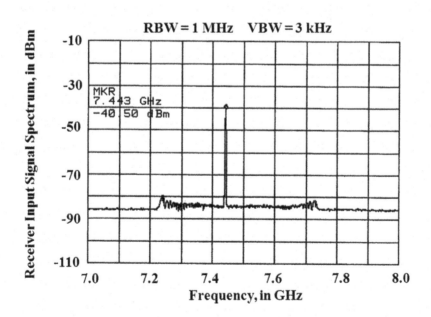

Figure 6.18 FM-UWB signal with CW interferer at 7.443 GHz.

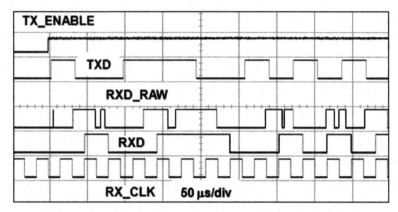

Figure 6.19   Measured receiver synchronization time at 31.25 kbps.

clock requires only a few bit transitions. According to [10], the mean time to have a well-aligned bit clock is below 5 bit transitions

6.4.6 Power Consumption

Table 6.2 presents a power consumption projection for the complete FM-UWB radio based upon the IC building blocks of this prototype and additional DDS and baseband electronics. All numbers are measured values, except for the DDS and baseband (BB) electronics consumption, which are simulated and estimated values for an implementation in 0.13 μm CMOS technology.

The measured power consumption of the RF VCO and output stage is 4 mW from a 1 V supply [5]. The measured power consumption of the receiver LNA and FM demodulator is 9 mW from a 1.8 V supply [7]. The measured power consumption of the subcarrier processor is 5.0 mW from a 1.8 V supply [9].

Table 6.2   Power consumption projection for the FM-UWB radio

Transmitter	5.5 mW
RF VCO	2.5 mW
RF output stage	1.5 mW
DDS & BB	1.5 mW
Receiver	**15.5 mW**
LNA	4.0 mW
FM demodulator	5.0 mW
Subcarrier processor	5.0 mW
DDS & BB	1.5 mW

The simulated power consumption of the DDS at 32 MHz clock frequency implemented in 0.13 μm CMOS technology is 400 μW and the simulated power consumption of the interpolating DAC is 350 μW [11]. A 32 MHz quartz oscillator with clock buffering is estimated to have a power consumption of 200 μW [12]. With 200 μW for some additional circuits, the total power consumption of the DDS plus baseband electronics is estimated to be 1.5 mW.

6.5 Conclusions

This chapter has presented measurement results made on IC building blocks and a complete hardware prototype of the FM-UWB radio operating at a center frequency of 7.45 GHz. The feasibility of the FM-UWB radio concept has been proven. The transmitter generates a FCC compliant signal and the receiver has a sensitivity of −85 dBm for a BER equal to 1×10^{-6}. This allows the implementation of a short-range radio link with 15 dB margin for a distance of 3 meters. The radio can operate with multiple FM-UWB users and also with narrowband interference. The fast receiver synchronization time has been confirmed.

Radio performance in a multi-user subcarrier FDMA scheme was found to be limited by sidelobe rejection of transmitter DDS and the receiver subcarrier channel selection filter.

Power consumption of a complete FM-UWB radio based upon the IC building blocks is estimated to be 5.5 mW in transmit mode and 15.5 mW in receive mode. These numbers are still relatively high, and there is room for improvement. Section 7.3 of this book shows directions for lowering the FM-UWB radio power consumption.

References

[1] IST-507102 project: Magnet Beyond – My personal Adaptive Global NET, http://www.magnet.aau.dk/.

[2] J.F.M. Gerrits, H. Bonakdar, M. Detratti, E. Pérez, M. Lobeira, Y. Zhao, Y. Dong, G. van Veenendaal, J.R. Long, J. R. Farserotu, E. Leroux and C. Hennemann, "A 7.2–7.7 GHz FM-UWB Transceiver Prototype," *Proc. ICUWB2009*, pp. 580–585.

[3] Nonin Medical Inc., *ipod Integrated Pulse Oximetry Device Specifications*, [Online]. Available: http://www.nonin.com/documents/ipod%20 Specifications.pdf.

[4] Y. Kim and D.-H. Kwon, "CPW-fed planar ultra wideband antenna having a frequency band notch function," *Electronics Letters*, vol. 40 no. 7, pp. 403–405, April 2004.

[5] Marco Detratti, Ernesto Perez, John F.M. Gerrits and Manuel Lobeira, "A 4.2 mW 6.25–8.25 GHz Transmitter IC for FM-UWB Applications," *Proc. ICUWB2009*, pp. 180–184.

[6] Agilent phase noise comparison.

[7] Yi Zhao, Yunzhi Dong, John F.M. Gerrits, Gerrit van Veenendaal, John R. Long and John R. Farserotu, "A Short Range, Low Data Rate, 7.2G-7.7GHz FM-UWB Receiver Front-End," *IEEE Journal of Solid-State Circuits*, vol. 44, no. 7, pp. 1872–1882, July 2009.

[8] P. Deixler, A. Rodriguez, W. de Boer, H. Sun, R. Colclaser, D. Bower, N. Bell, A. Yao, R. Brock, Y. Bouttement, G.A.M. Hurkx, L.F. Tiemeijer, J.C.J. Paasschens, H.G.A. Huizing, D.M.H. Hartskeerl, P. Agarwal, P.H.C. Magnee, E. Aksen and J.W. Slotboom, "QUBiC4X: An f_T/f_{max} = 130/140GHz SiGe:C-BiCMOS Manufacturing Technology with Elite Passives for Emerging Microwave Applications," *Proc. BCTM 2004*, pp. 233–236.

[9] Hamid Bonakdar, John Gerrits, Dominique Noguet, Gerrit van Veenendaal, Yi Zhao, Magnet Beyond deliverable D5.4.2b, *LDR High Band Test Report*, September 2008.

[10] Helmuth Brügel, and Peter F. Driessen, "Variable Bandwidth DPLL Bit Synchronizer with Rapid Acquisition Implemented as a Finite State Machine," *IEEE Transactions on Communications*, vol. 42, no. 9, pp. 2751–2759, September 1994.

[11] Private communication from Peter Nilsson, Lund University.

[12] Private communication from Erwan Leroux, CSEM.

7

Conclusions

This book has presented the FM-UWB communication system, illustrated its performance, and provided hardware implementation examples. This chapter presents findings, a summary of the key contributions and recommendations for future work.

7.1 Main Findings

New wireless applications such as health monitoring and body-area networks (BAN) require tetherless connectivity at data rates below 250 kbps, a range less than 10 m, and operational lifetime from a single battery charge of weeks or months. The LDR radio for such applications needs to be low-complexity and yet robust to interference and frequency-selective multipath and also be able to rapidly join or leave a network. The short PHY synchronization time simplifies the task of the medium access control (MAC), which would then be similar to the ones used in low-power narrowband, but not necessarily robust, radios.

This book has explored the constant-envelope frequency modulation ultra-wideband (FM-UWB) radio that uses a low-complexity implementation based upon a low-frequency FSK subcarrier modulation followed by wideband analog FM to spread the spectrum. The subcarriers can be seen as analog spreading codes. Different users may use subcarrier FDMA to share the same RF spectrum and distinguish themselves by their subcarrier frequency.

Instantaneous despreading in the receiver is realized by an analog FM demodulator. Real processing gain is obtained by the bandwidth reduction after the FM demodulator. Due to instantaneous despreading in the wideband FM demodulator, the system behaves like a narrowband FSK system where synchronization is limited only by the bit synchronization time.

It was shown that the triangular subcarrier waveform yields best performance in terms of flatness of the RF power spectral density and rapid roll-off

151

of the FM-UWB signal. Constant-envelope subcarrier modulation schemes, like 2-FSK or 4-FSK are required to obtain a flat power spectral density. The optimum subcarrier frequency lies between 1 and 2 MHz. Higher subcarrier frequencies will require reduction of the transmit power in order to comply with the emission limits.

The fixed time delay FM demodulator is the best candidate for simultaneous demodulation of multiple FM-UWB signals. It is best used with an overdrive equal to unity, i.e., the demodulator bandwidth equals the FM-UWB signal bandwidth. The FM-UWB approach is tolerant to frequency offset between the transmitter and receiver. A 1 dB sensitivity reduction occurs for ±100 MHz offset in the center frequency for a demodulator bandwidth of 500 MHz.

Performance with AWGN shows a link span that is well suited to short-range BAN applications. Up to 17 meters is obtained under free-space propagation conditions for a 500 MHz wide FM-UWB signal at 250 kbps and assuming a 5 dB noise figure for the receiver LNA.

The performance with frequency-selective multipath (typically the IEEE UWB and BAN channel models) has been investigated. The mean and median values of the fading distribution were both found to be 0 dB, meaning that 50% of the time performance improves and 50% of the time performance degrades. For the IEEE802.15.6 BAN channels, it was found that 2.8 dB of fading margin is required to achieve 99% availability in the most difficult propagation conditions. This compares favorably to a narrowband radio, which requires 20 dB higher received power for 99% availability.

Multiple-access interference is generated in the wideband FM demodulator and attenuated by the receiver processing gain, which is equal to the ratio of the RF and subcarrier bandwidths. For example, in a 100 kbps system with a RF bandwidth of 500 MHz, a processing gain of 34 dB is available. This allows for a FM-UWB interferer that is 24 dB stronger than the wanted signal, a value that is realistic for a short-range LDR BAN scenario.

The maximum number of users in a FM-UWB communication system based upon subcarrier FDMA is dependent on the bit rate. In theory, a 10 kbps system with a RF bandwidth of 1 GHz can accommodate 50 users at equal power levels, whereas a 100 kbps system accommodates 15 users, provided that the subcarrier frequency is chosen sufficiently high. It was shown in Chapter 2 that the use of subcarrier frequencies above 2 MHz requires a reduction in the transmit power, which lowers the link span. A trade-off needs to be made between the number of simultaneous users and the link span. The result may be that the subcarrier frequency range is restrained.

The capacity of a FM-UWB system using binary FSK as subcarrier modulation scheme is equal to one half of the available subcarrier frequency range. When the optimum subcarrier frequency range (i.e., 1–2 MHz) is used, this yields a capacity of 500 kbps. The maximum capacity cannot be exploited without backing off transmit power.

The effect of in-band interference is mitigated by the receiver processing gain. In-band interference whose bandwidth is smaller than the FM-UWB signal bandwidth can be attenuated by a notch filter. The effect of such a notch upon the receiver sensitivity is negligible. Out-of-band interference can be reduced significantly by passive filtering prior to the UWB receiver. Antennas can also be designed to have notches in their frequency response.

The implementation of the FM-UWB radio has also been investigated and the influence of hardware imperfections like oscillator phase noise and circuit non-linearities on the performance of the radio has been quantified.

Direct digital synthesis (DDS) techniques are used for subcarrier generation in the transmitter, yielding a flexible design with low parts count in hardware and accuracy in terms of subcarrier frequency and deviation. Analog and digital techniques are combined for the RF part. The FM-UWB signal is generated by a free-running RF VCO that is calibrated by a PLL frequency synthesizer. VCO tuning curve non-linearity affects the PSD of the FM-UWB signal. When we allow for 1 dB of peaking of the PSD of the FM-UWB signal, the VCO gain at the band edges should not be lower than 80% of the value at the VCO center frequency.

Transmitter phase noise limits the receiver subcarrier SNR in a similar way as multiple-access interference. To be multiple-access limited in a 250 kbps system, it was shown that the transmitter phase noise needs to be lower than −80 dBc/Hz at 1 MHz offset from the carrier. Such an oscillator can be realized at low resonator power ($<1\,\mu W$) when a harmonic oscillator with resonator quality factor of 10 is used. Power consumption of the oscillator will be limited by parasitic capacitances. Current reuse techniques may be used to bias the oscillator and output amplifier.

The FM-UWB receiver performance is strongly influenced by the wideband FM demodulator. The difficulties are in the implementation of a flexible (i.e., tunable) time delay circuit and a high-gain, low-noise multiplier circuit.

The tunable delay circuit is advantageously implemented by a cascade of a lattice bandpass filter that yields −90 degrees of phase shift at its resonant frequency, and a simple bandpass filter. The double-balanced Gilbert multiplier makes a good choice for the multiplier circuit and its noise can be lowered by current bleeding, which yields a sensitivity improvement. The diverted current can then be reused to bias gain stages in the delay circuit and the LNA.

Adding preamplification to the delay circuit of the FM demodulator lowers the gain compression point more than the sensitivity is increased, yielding reduced dynamic range. The gain in the delay circuit should therefore be chosen equal to unity. Adding gain in the LNA, or adding an intermediate gain stage between the LNA and FM demodulator improves the sensitivity with a minimum penalty on the large-signal behavior. Automatic gain control (AGC) may be applied to further increase the dynamic range.

The subcarrier processor uses a direct-conversion approach. However, due to the presence of harmonics of the subcarriers at the FM demodulator output, additional filtering is required prior to downconversion to baseband of the FSK subcarrier signal. This is performed by an analog lowpass filter. Additional attenuation of the odd harmonics by 30 dB is obtained by using a triangular LO signal in the downconversion mixers.

The FM-UWB radio implementation phase was completed by construction and characterization of a hardware prototype operating at a center frequency of 7.45 GHz using dedicated ICs and off-the-shelf components. The prototype was also used for successful FCC pre-compliance testing.

7.2 Original Contributions

This section summarizes the key contributions of this work:

This book has presented for the first time a UWB communication system based upon constant-envelope modulation. FM-UWB radio exploits a simple but effective FM scheme for spreading a narrowband FSK signal and thus obtaining an ultra-wideband signal compliant with regulations. The FM-UWB signal is created by modulating the RF VCO with the triangular FSK subcarrier. This results in a UWB signal with flat power spectral density and steep spectal roll-off. The despreading in the receiver is implemented a fixed time delay FM demodulator. Since despreading is instantaneous, the FM-UWB radio behaves like a narrowband FSK radio where synchronization is only limited by the receiver bit synchronization.

The use of double modulation, that is, digital FSK followed by analog FM used to generate a UWB signal and to transmit data, was first published in 2003. This approach yields a low-complexity implementation of a spread-spectrum system. The users may share the same RF spectrum and distinguish themselves by their subcarrier frequency. The subcarriers can be seen as analog spreading codes.

True processing gain is available due to the bandwidth reduction after the receiver FM demodulator. The processing gain is active against interference from other in-band users. The analog spread-spectrum approach yields robustness against frequency-selective multipath. It also allows mitigation of narrowband envelope-modulated in-band interference by notch filtering prior to the FM demodulator with low impact on the receiver performance.

The performance of the FM-UWB radio with frequency selective multipath has been modeled in an efficient way. The frequency domain transfer function of the channel over the receiver bandwidth, calculated from the channel impulse response, suffices to evaluate the impact on the subcarrier SNR. No long simulations are required to obtain statistics for the various propagation channels. The influence of filtering in the receiver front-end can be modeled in the same way.

A complete transceiver prototype that confirms theoretical results has been developed. Circuit implementations based upon current reuse techniques have demonstrated the low power consumption potential of the FM-UWB radio. DDS techniques yield a low-complexity solution to generate FSK and local oscillator signals with accurate control of parameters. The use of triangular LO signal in the receiver subcarrier processor further helps to lower power consumption.

The feasibility of the FM-UWB concept was demonstrated in a hardware prototype operating at a center frequency of 7.45 GHz in 2008. FM-UWB was proposed as one of the physical layers for the IEEE 802.15.6 BAN standard and accepted as optional mode.

7.3 Recommendations for Future Work

This work has presented the properties of the FM-UWB communication system and demonstrated the feasibility of a low-complexity implementation of the FM-UWB radio. Further investigations on worst case propagation conditions (3.3) and the influence of narrowband interference (3.4) are proposed for a deeper understanding of the theoretical limitations. Work this also out in Chapter 3 and in 6 for CW inteferer.

Lessons have been learned about how to improve performance and lower power consumption. A full CMOS integration of the FM-UWB transceiver would push the limits of power consumption. Power consumption targets would be on the order of 1 mW for the transmitter and 5 mW for the receiver. As an example, a receiver front-end with current bleeding in the FM demodulator, where part of the reused current serves in an additional gain stage between LNA and FM demodulator, would be worthwhile to explore.

Packaging the FM-UWB radio together with its antenna and micro-controller running a low power MAC protocol. and obtaining a small form factor.

An additional dimension would be to investigate a FM-UWB system using a 4-FSK subcarrier modulation scheme to increase transmit rate. This would require a more sophisticated FSK demodulator based upon, e.g., a phase digitizer in the receiver.

Finally, it could be advantageous to combine the FM-UWB radio with a narrowband FSK radio (e.g., operating at 2.4 or 5.8 GHz). The low-power FM-UWB transmitter is used for the transmission of short-range telemetry data. The narrowband FSK radio may be used for applications that require increased range. Both radios could be combined on the same die and share the FSK circuits and use separate RF front-ends.

Appendix A

Power Spectral Density for FSK and BPSK Subcarrier Modulation Schemes

A.1 Constant-envelope Subcarrier Modulation Scheme

Figure A.1 shows raw transmit data, digitally lowpass filtered data and the 2-FSK subcarrier signal in the time domain. The FSK subcarrier signal is constant-envelope. Figure A.2 shows the probability density function of the subcarrier signal and the normalized power spectral density of the FM-UWB signal. The subcarrier signal's PDF is uniform which is reflected in the PSD of the FM-UWB signal which is flat [1].

A.2 Non Constant-envelope Subcarrier Modulation Scheme

The use of pulse amplitude modulation (PAM) or phase modulation (PM) for the subcarrier modulation scheme is not recommended. Figure A.3 shows raw transmit data, digitally lowpass filtered data and the BPSK subcarrier signal in the time domain. The BPSK subcarrier signal is not constant-envelope. Figure A.4 shows the PDF of the subcarrier signal and the normalized power spectral density of the FM-UWB signal. The subcarrier signal's PDF is not uniform which is reflected in the PSD of the FM-UWB signal which is not flat at all. Use of the BPSK subcarrier modulation scheme would require considerable back-off, approximately 8 dB, on the transmitter side to respect the UWB spectral mask.

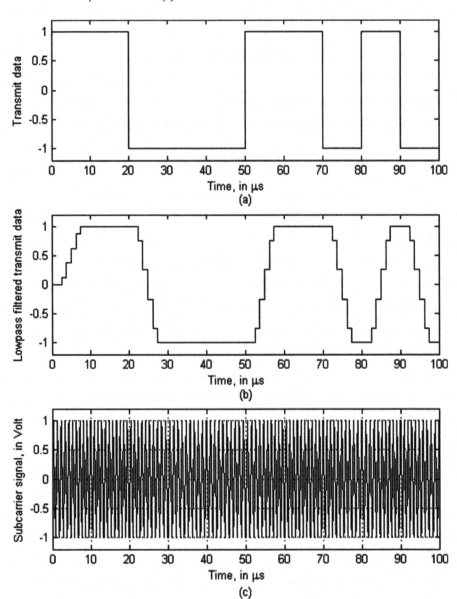

Figure A.1   (a) Transmit data, (b) lowpass filtered transmit data, and (c) FSK subcarrier signal.

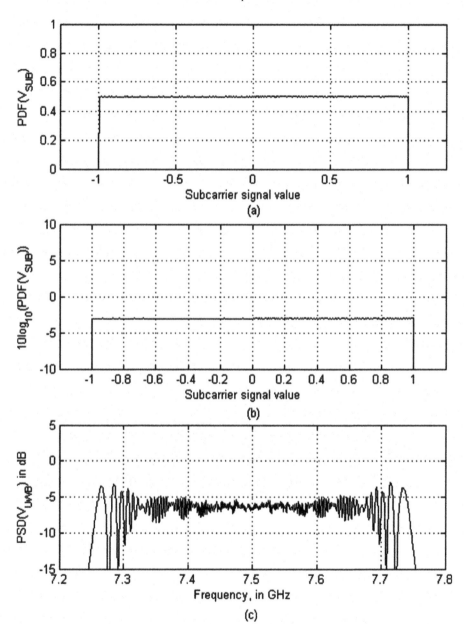

Figure A.2 (a) PDF of subcarrier on linear scale, (b) PDF of subcarrier on logartithmic scale, (c) power spectral density of FM-UWB signal using 2-FSK subcarrier modulation, resolution bandwidth equals 1 MHz.

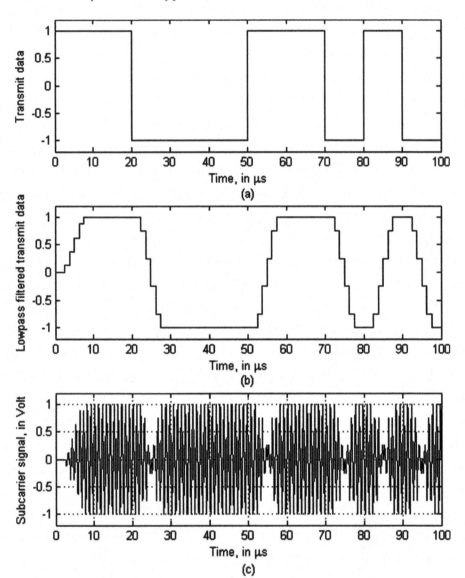

Figure A.3   (a) Transmit data, (b) lowpass filtered tranmit data, and (c) BPSK subcarrier signal.

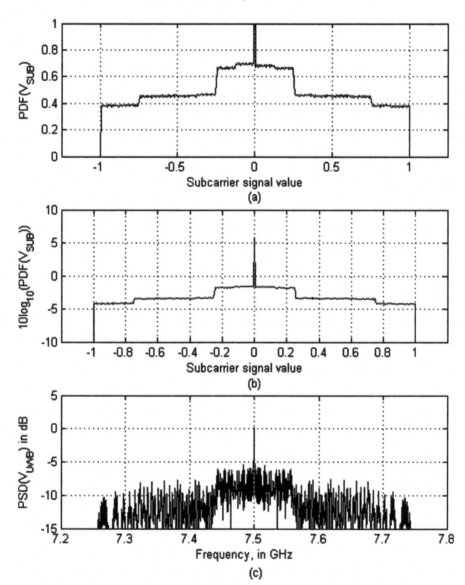

Figure A.4 (a) PDF of subcarrier on linear scale, (b) PDF of subcarrier on logarithmic scale, (c) power spectral density of FM-UWB signal using BPSK subcarrier modulation, resolution bandwidth equals 1 MHz.

Reference

[1] Nelson M. Blachman, George A. McAlpine, "The Spectrum of a High-Index FM Waveform: Woodward's Theorem Revisited," IEEE Transactions on Communication Technology, volume 17, issue 2, pp. 201–208, April 1969.

Appendix B

Influence of RF Frequency Notch on Subcarrier Level

This appendix shows the effect of notch filtering on the level of the fundamental of the subcarrier signal. A notch in the center of the FM-UWB signal has minimum effect.

B.1 Notch at Center Frequency with Variable Width

This section shows simulation results for a frequency notch placed in the center of the bandwidth occupied by the FM-UWB signal. Simulation results have been obtained for a brick wall notch filter with bandwidth B_{NOTCH}. The notch width Δ is defined as

$$\Delta = \frac{B_{NOTCH}}{B_{RF}} \tag{B.1}$$

Figure B.1 illustrates the case of a 40% wide notch ($\Delta = 0.4$) in the center of the FM-UWB signal.

B.2 Sliding Notch

The influence of a sliding brick wall notch of fixed 100 MHz bandwidth passing through the FM-UWB signal is illustrated in Figures B.3 and B.4. Such a notch could be the results of a notch in the LNA meant to attenuate, e.g., WiMAX signals. The FM-UWB signal is 500 MHz wide and has a center frequency of 7.5 GHz. The notch center frequency is varied from 7.3 up to 7.7 GHz. When the notch is at the edges of the FM-UWB signal, the fundamental of the subcarrier goes down by 4 dB, corresponding to a 2 dB sensitivity loss at RF.

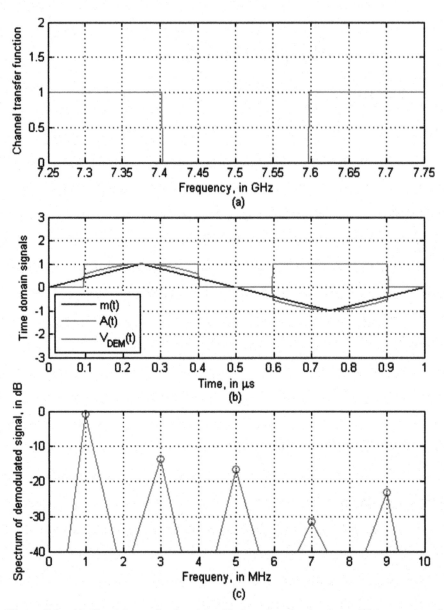

Figure B.1   (a) Channel transfer function, (b) time domain signals, and (c) spectrum of FM demodulator output signal for a notch whose width equals 40% of the FM-UWB signal bandwidth.

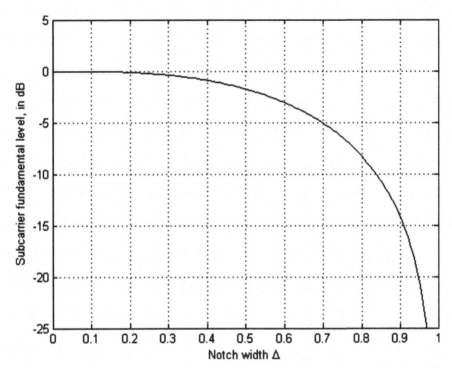

Figure B.2 Subcarrier level versus notch width $\Delta = B_{NOTCH}/B_{RF}$. for a notch in the middle of the FM-UWB signal.

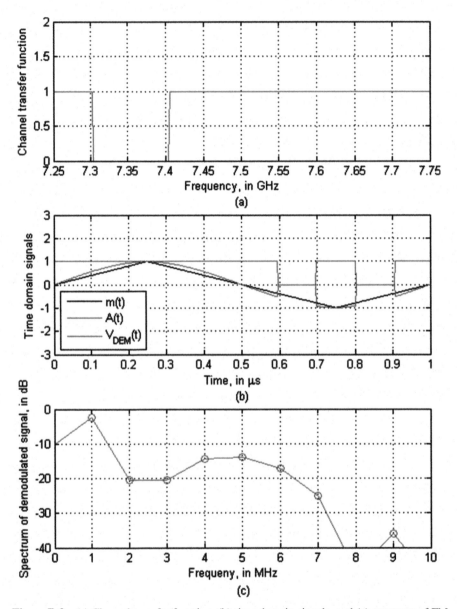

Figure B.3   (a) Channel transfer function, (b) time domain signals, and (c) spectrum of FM demodulator output signal for a 100 MHz wide notch ($\Delta = 0.2$), 150 MHz away from the center of the FM-UWB signal.

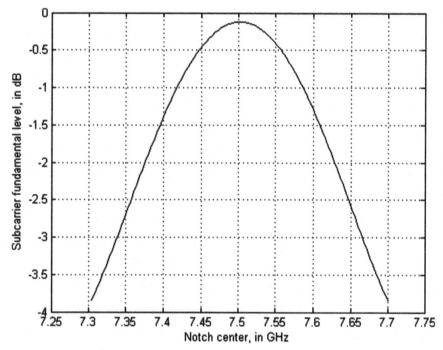

Figure B.4   Subcarrier level versus notch center frequency for a 100 MHz wide notch.

Appendix C

Detailed Schematics of Wideband FM Demodulator

This appendix provides detailed schematics of the receiver front-end described in [1]. It comprises the low noise amplifier and FM demodulator that are implemented in NXP Semiconductors' QUBiC4X 0.25 μm SiGe:C-BiCMOS technology [2]. Target power consumption of the receiver front-end demonstrator was 10mW from a 1.8 V supply. The measured consumption is 2 mA in the LNA, and 3.1 mA in FM demodulator, yielding a total of 5.1 mA from a 1.8 V supply, that is 9.2 mW.

C.1 FM Demodulator

The FM demodulator described in [1] uses a Gilbert multiplier constituted by transistors Q_9 to Q_{16}. The delay circuit comprises two stages. A First gain stage (trans-conductance $g_{m1} = 14$ mA/V) constituted by differential pair Q_1–Q_2 and Miller compensation transistors Q_3 and Q_4, drives the LBPF resonator with component values $R_{APF} = 12\ \Omega$, $L_{APF} = 2.5$ nH, and $C_{APF} = 180$ fF.

The second gain stage (trans-conductance $g_{m2} = 14$ mA/V) constituted by differential pair Q_5–Q_6 and Miller compensation transistors Q_7 and Q_8, drives the BPF resonator, which is connected to the quad of the multiplier. Component values are $R_{APF} = 12\ \Omega$, $L_{BPF} = 1.0$ nH, and $C_{APF} = 440$ fF.

C.2 Low Noise Amplifier

The LNA has a single-ended input and is designed to operate with a 50 Ω antenna and prefilter. Gain control with approximately 10 dB range is also included. Figure C.2 shows the LNA schematic. The Miller effect due to the collector-base capacitance of transistors Q_1–Q_4 is neutralized by feedback capacitors implemented using transistors Q_5–Q_8, which have the same base-collector capacitance as transistors Q_1–Q_4.

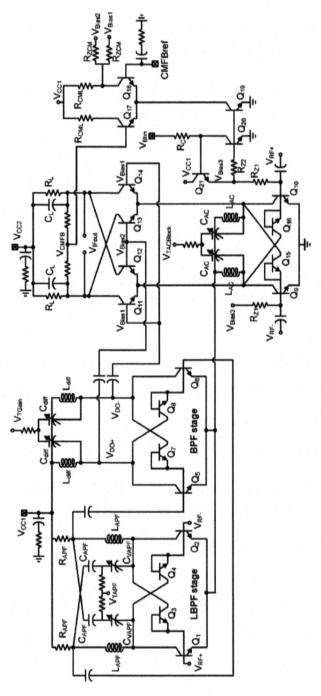

Figure C.1   FM demodulator schematic [1].

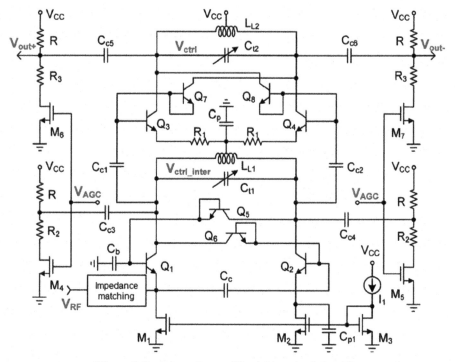

Figure C.2 Low noise amplifier with active balun [1].

References

[1] Y. Zhao, Y. Dong, J.F.M. Gerrits, G. van Veenendaal, J.R. Long and J.R. Farserotu, "A Short Range, Low Data Rate, 7.2G–7.7 GHz FM-UWB Receiver Front-End," *IEEE Journal of Solid-State Circuits*, vol. 44, no. 7, pp. 1872–1882, July 2009.

[2] P. Deixler, A. Rodriguez, W. de Boer, H. Sun, R. Colclaser, D. Bower, N. Bell, A. Yao, R. Brock, Y. Bouttement, G.A.M. Hurkx, L.F. Tiemeijer, J.C.J. Paasschens, H.G.A. Huizing, D.M.H. Hartskeerl, P. Agarwal, P.H.C. Magnee, E. Aksen and J.W. Slotboom, "QUBiC4X: An f_T/f_{max} = 130/140 GHz SiGe:C-BiCMOS Manufacturing Technology with Elite Passives for Emerging Microwave Applications," *Proc. BCTM 2004*, pp. 233–236.

Appendix D

Approximation of Small-signal Gain of Hyperbolic Tangent Non-linearity

The average small-signal gain of the hyperbolic tangent non-linearity $y = \tanh(x)$ for an overdrive O is given by

$$A_{v,avg} = \int_0^{2\pi} \left(1 - \tanh^2\left(O\sin x\right)\right) dx \tag{D.1}$$

This integral doesn't have an explicit solution, but can be evaluated numerically. For further calculations, an explicit expression Av,avg(O) was required. The integral was approximated by the weighted sum of two approximations $A_{v1,avg}$ and $A_{v2,avg}$:

$$A_{v1,avg} = \exp\left(-0.4O^{1.4}\right) \tag{D.2}$$

$$A_{v2,avg} = \frac{1}{1 + \frac{O}{0.77}} \tag{D.3}$$

$$A_{v,avg} = W_1 A_{v1,avg} + W_2 A_{v2,avg}$$

$$= 0.982 \left[\frac{1}{1 + \frac{O}{1.8}} A_{v1,avg} + \frac{O}{1 + \frac{O}{1.12}} A_{v2,avg}\right] \tag{D.4}$$

The first approximation works well for values O < 2 and the second approximation for values O > 4. The weighing functions W_1 and W_2 can be interpreted as lowpass and highpass transfer functions.

Figure D.1 shows that the approximation error is less than 0.5 dB for overdrive values O < 10.

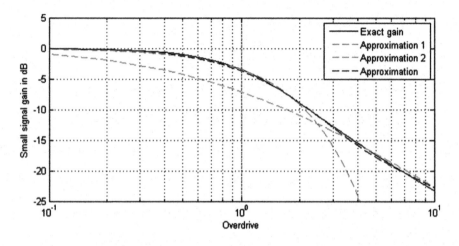

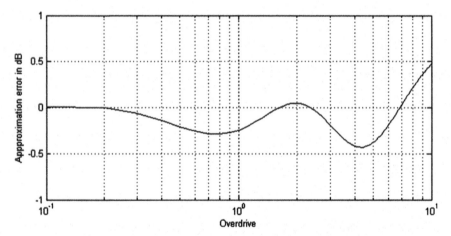

Figure D.1   Approximation of small-signal gain.

List of Publications

Journal Papers and Book Chapters

J. R. Farserotu, **J. F. M. Gerrits** and J. Rousselot, "Low power and robust PHY-MAC solution for Medical BAN", IEICE Transactions on Communications, Vol. E93-B, No. 4, pp. 802–810, April 2010.

Y. Zhao, Y. Dong, **J. F. M. Gerrits**, G. van Veenendaal, J. R. Long and J. R. Farserotu, "A Short Range, Low Data Rate, 7.2G–7.7GHz FM-UWB Receiver Front-End," IEEE J. Solid-State Circuits, Vol. 44, No. 7, pp. 1872–1882, July 2009.

J. F. M. Gerrits, J. R. Farserotu and J. R. Long, "Low-Complexity Ultra Wideband Communications," IEEE Transactions on Circuits and Systems-II, Vol. 55, No. 4, pp. 329–333, April 2008.

Z. Shelby, J. Farserotu and **J. F. M. Gerrits**, "Small, Cheap Devices for Wirelesss Sensor Networks", in Subhir Dixit, Ramjee Prasad, Editors, "Technologies for Home Networking," Hoboken: J. Wiley & Sons, ISBN: 978-0470073742, 2008.

J. F. M. Gerrits, M. H. L. Kouwenhoven, P. R. van der Meer, J. R. Farserotu and J. R. Long, "Principles and Limitations of Ultra Wideband FM Communications Systems," EURASIP Journal on Applied Signal Processing, Volume 2005, Number 3, pp. 382–396, March 2005.

Conference Papers

J. F. M. Gerrits, M. Danesh, Y. Zhao, Y. Dong, G. van Veenendaal, J. R. Long, and J. R. Farserotu, "System and Circuit Considerations for Low-Complexity Constant-Envelope FM-UWB", Proceedings ISCAS 2010, pp. 3300–3303, June 2010.

J. F. M. Gerrits, H. Bonakdar, M. Detratti, E. Pérez, M. Lobeira, Y. Zhao, Y. Dong, G. van Veenendaal, J. R. Long, J. R. Farserotu, E. Leroux and C. Hennemann, "A 7.2–7.7 GHz FM-UWB Transceiver Prototype," Proceedings ICUWB 2009, pp. 580–585, September 2009.

M. Detratti, E. Perez, **J. F. M. Gerrits** and M. Lobeira, "A 4.2 mW 6.25–8.25 GHz Transmitter IC for FM-UWB Applications," Proceedings ICUWB 2009, pp. 180–184, September 2009.

Y. Zhao, G. van Veenendaal, H. Bonakdar, **J. F. M. Gerrits** and J. R. Long, "3.6mW 30dB Gain Preamplifiers for FM-UWB Receiver," Proceedings BCTM 2008, pp. 216–219, October 2008.

J. R. Long, W. Wu, Y. Dong, Y. Zhao, M. A. T. Sanduleanu, **J. F. M. Gerrits** and G. van Veenendaal, "Energy-efficient Wireless Front-end Concepts for Ultra Lower Power Radio," Proceedings CICC 2008, pp. 587–590, September 2008.

Y. Dong, Y. Zhao, **J. F. M. Gerrits**, G. van Veenendaal and J. R. Long, "A 9 mW High Band FM-UWB Receiver Front-end," Proceedings ESSCIRC 2008, pp. 302–305, September 2008.

J. F. M. Gerrits, J. R. Farserotu and J. R. Long, "Multipath Behavior of FM-UWB Signals," Proceedings ICUWB 2007, pp. 162–167, September 2007.

J. F. M. Gerrits, J. R. Farserotu and J. R. Long, "Low-Complexity Ultra Wideband Communications," Proceedings ISCAS 2007, pp. 757–760, May 2007.

P. Nilsson, **J. F. M. Gerrits**, J. Yuan, "A Low Complexity DDS IC for FM-UWB Applications," Proceedings IST Mobile & Wireless Communications Summit 2007, pp. 1–5, July 2007.

J. F. M. Gerrits, J. R. Farserotu and J. R. Long, "A Wideband FM Demodulator for a Low-Complexity FM-UWB Receiver," Proceedings ECWT 2006, pp. 99–102, September 2006.

J. F. M. Gerrits, J. R. Farserotu and J. R. Long, "A Wideband FM Demodulator for UWB Applications," Proceedings PRIME 2006, pp. 461–464, June 2006.

J. F. M. Gerrits, J. R. Farserotu and J. R. Long, "Multiple-user Capabilities of FM-UWB Communications Systems," Proceedings ICU 2005, pp. 684–689, September 2005.

J. F .M. Gerrits, J. R. Farserotu and J. R. Long, "Multiple-Access Interference in FM-UWB Communication Systems," Proceedings WPMC 2005, pp. 2027–2031, September 2005.

J. F. M. Gerrits, J. R. Farserotu and J. R. Long, "UWB Considerations for "My Personal Global Adaptive Network" (MAGNET) Systems," Proceedings ESSCIRC 2004, pp. 45–56, September 2004.

J. F. M. Gerrits and J. R. Farserotu, "Ultra Wide Band FM: A Constant Envelope Frequency Domain Approach," Proceedings IZS 2004, pp. 90–93, February 2004.

J. F. M. Gerrits and J. R. Farserotu,"Ultra Wideband FM: A Straightforward Frequency Domain Approach," Proceedings EUMW 2003, pp. 853–856, October 2003.

Summary

This book describes the principles, performance and practical implementation of constant-envelope frequency modulation ultra-wideband (FM-UWB) communication systems.

New wireless applications such as health monitoring and body-area networks (BAN) require tetherless connectivity at data rates below 250 kbps, a range less than 10 m, and low power consumption. The LDR radio for such applications needs to be low-complexity and yet robust to interference and frequency-selective multipath and also be able to rapidly join or leave a network. The short PHY synchronization time simplifies the task of the medium access control (MAC), which can be similar to the ones used in low-power narrowband, but not necessarily robust, radios. The challenge is to implement a low-complexity, compact, low-power spread spectrum radio with instantaneous despreading in the receiver (no pulse or sequence synchronization).

Chapter 2 presents the constant-envelope frequency modulation ultra-wideband (FM-UWB) approach, which uses a low-complexity implementation based upon a low-frequency FSK subcarrier followed by wideband analog FM to implement spread spectrum. The use of a triangular subcarrier waveform and FSK subcarrier modulation yields a flat power spectral density and steep spectral roll-off. In the receiver, the instantaneous despreading is implemented by a wideband FM demodulator. FM-UWB is tolerant to offset between transmit and receive center frequencies.

Chapter 3 presents the performance of the FM-UWB communication system. With AWGN, the FM-UWB radio requires approximately 15 dB more energy per bit compared to a narrowband FSK system, however, in return the FM-UWB approach provides robustness against interference and multipath that is not available in narrowband FSK systems. The system provides a good link margin for short-range LDR BAN applications. The performance with frequency-selective multipath, typically the IEEE UWB and BAN channel models, has been investigated. The mean and median values of the fading distribution are both found to be 0 dB, meaning that

50% of the time performance improves and 50% of the time performance degrades. With frequency-selective fading as modeled by the IEEE 802.15.6 BAN channels, 2.8 dB of fading margin is required to achieve 99% availability in the worst case BAN channel. This compares favorably to a narrowband radio, which requires 20 dB higher received power for 99% availability. The effect of in-band interference is mitigated by the receiver processing gain, which is equal to the ratio of the RF and subcarrier bandwidth. In-band interference whose bandwidth is smaller than the FM-UWB signal bandwidth can be attenuated by a notch filter. The effect of such a notch upon the receiver sensitivity is negligible. Out-of-band interference can be reduced significantly by passive filtering prior to the UWB receiver. Antennas can also be designed to have notches in their frequency response. Multiple users may be accommodated in a number of ways. Apart from standard TDMA and RF FDMA techniques, FM-UWB may also use subcarrier FDMA techniques by assigning different subcarrier frequencies to different users. By avoiding hardlimiting in the receiver, simultaneous demodulation of multiple FM-UWB input signals with different subcarrier frequencies occupying the same RF band is possible. The number of simultaneous users is limited by multiple access interference and the processing gain. Apart from this fundamental limitation, the performance in a subcarrier FDMA system may be limited by implementation-dependent hardware imperfections such as insufficient sidelobe rejection in the transmitter subcarrier generation, transmitter RF oscillator phase noise, and insufficient selectivity of the receiver subcarrier filtering.

Chapter 4 addresses the transmitter implementation. Direct digital synthesis (DDS) techniques are used for subcarrier generation, yielding a flexible design with low parts count in hardware and accurate output signals in terms of frequency and deviation. The FM-UWB signal is generated by a free-running RF VCO that can be calibrated by a PLL frequency synthesizer. Relaxed hardware specifications in terms of VCO phase noise (–80 dBc/Hz at 1 MHz offset), enable low power consumption of the transmitter.

Chapter 5 considers details of the receiver implementation. The performance of the FM-UWB receiver is strongly influenced by the wideband FM demodulator. The difficulties are in the implementation of a flexible (i.e., tunable) time delay circuit and a high-gain, low-noise multiplier circuit. The tunable delay circuit is advantageously implemented by a cascade of a lattice bandpass filter that yields –90 degrees of phase shift at its resonant frequency, and a simple bandpass filter. The double-balanced Gilbert multiplier makes a good choice for the multiplier circuit and its noise can be lowered by current bleeding, which yields a sensitivity improvement. The diverted current can then be reused to bias gain stages in the delay circuit and the LNA. The gain in the delay circuit should therefore be chosen equal to unity. Adding gain in the LNA, or adding an intermediate amplifier between the LNA and FM demodulator improves the sensitivity with a minimum penalty on the large-signal behavior. The subcarrier processor uses a direct-conversion approach. Due to the presence of harmonics of the subcarriers at the FM demodulator output, additional filtering is required prior to downconversion to baseband of the FSK subcarrier signal. This is performed by an analog lowpass filter. Additional attenuation of the odd harmonics by 30 dB is obtained by using a triangular LO signal in the downconversion mixers.

Chapter 6 presents measurement results made on IC building blocks and a complete hardware prototype of the FM-UWB radio operating at a center frequency of 7.45 GHz. The feasibility of the FM-UWB radio concept has been proven. The transmitter generates a FCC compliant signal and the receiver has a sensitivity of –85 dBm for a BER equal to 1×10^{-6}. This allows the implementation of a short-range radio link with 15 dB margin for a distance of 3 meters. The radio can operate with multiple FM-UWB users and also with narrowband interference. Radio performance in a multi-user subcarrier FDMA scheme was found to be limited by the sidelobe rejection of the transmitter DDS and the receiver subcarrier channel selection filter. The fast receiver synchronization time has been confirmed. Power consumption of a complete FM-UWB radio based upon the IC building blocks is estimated to be 6 mW in transmit mode and 16 mW in receive mode. These numbers are still relatively high, and there is room for improvement.

Index

About the Author

John F. M. Gerrits was born in 1963 in Leiden, the Netherlands. He received the M.Sc.EE. degree from Delft University of Technology, the Netherlands, in 1987. His final book was on the design of integrated high-performance harmonic oscillator circuits. In 1988 he joined the Philips T&M division in Enschede, the Netherlands, where he designed integrated oscillator and data-acquisition circuits for oscilloscope applications.

In 1991 he joined CSEM in Neuchâtel, Switzerland, where he has been involved in both system and circuit design of a single-chip low-power VHF radio receiver for hearing aid applications and of a single-chip ISM UHF transceiver. His current work involves system and circuit design of UWB radio systems, RF and microwave sensing techniques and measurement methodology.

He is editor and co-author of the book Low-power Design Techniques and CAD tools for analog and RF integrated circuits published by Kluwer, in 2001. He holds 3 European and 5 US patents and is the winner of the 2006 European Conference on Wireless Technologies Prize.

John Gerrits passed away in February 2011.

Lightning Source UK Ltd.
Milton Keynes UK
UKOW06n1146270117
293027UK00002B/17/P